MEMORIAL

OF

COLONEL SAMUEL COLT.

THE following tribute to the inventive genius of Col. SAMUEL COLT, with a record of the wonderful change wrought in the Hartford South Meadow by the construction of his dyke, armory, and homestead, was projected immediately after his death, in 1862. The earlier completion of this memorial has been delayed, partly by the pressure of prior engagements, and partly by its enlargement to make it more worthy of the costly illustrations promptly furnished by Mrs. Colt, when made acquainted with the design. Its publication now, in the midst of new engagements quite foreign to the work of composition, is made possible only through the hearty coöperation of Prof. J. D. Butler, of the State University of Wisconsin, who had in his own felicitous manner described the Colt revolver, and its manufacture, after a visit to the armory in 1863. The Memoir, originally drawn up to be used only as material for this purpose, is contributed by one who alone could speak from the fulness of personal knowledge, and the tenderness of personal bereavement, especially of the domestic side of his life and character. To the whole is applied the designation of ARMSMEAR—by which the Colt estate, and especially the homestead, is known—on account of the change wrought by his ARM in the *mear*, by which, in our old Saxon speech, this meadow would be designated.

HENRY BARNARD.

ST. JOHN'S COLLEGE, *Annapolis*, Maryland.

Ever faithfully Yours
Saml Colt

ARMSMEAR:

THE HOME,

THE ARM, AND THE ARMORY

OF

SAMUEL COLT.

A Memorial.

NEW YORK.

M.DCCC.LXVI.

CONTENTS.

_____ ___._____ ____

2

CONTENTS.

LIST OF ILLUSTRATIONS.

ARMSMEAR

The House
OR
The Armory, the Field

I. ARMSMEAR.

TANDING on the brow of Wyllys
Hill—as that gentle eminence was
called till quite recently, but now
known as Charter Oak Place, on the
terrace which extends north and south
from Little River, and rises as it recedes
to the west into the beautiful uplands

on which the town of Hartford is built, but which once
doubtless formed the barriers of broader and deeper waters than now flow between the present banks of Connecticut River—we have, or could have had, ten years ago,
before that historic hill was appropriated to private
dwellings, a full view of that portion of South Meadow
which the first settlers of the colony, both Dutch and
English, there is good reason to believe, designed to
occupy permanently with streets, and structures for business and residence. On the left bank of Little River,
the northern boundary of the meadow, at its junction
with Great or Connecticut River, the Dutch, in 1614,
made their first landing, and in 1633 erected the *House
of Hope*, on land purchased from the Indian sachem Ne-
paquash, who had his hunting-grounds, and fishery, and
cornfields, as well as the burial-place of his tribe, here-
about. In 1636, those lights of the Western churches,
the Rev. Thomas Hooker, Samuel Stone, John Haynes,
Thomas Welles, and their associates, and, two years later,
George Wyllys and Edward Hopkins, having purchased
the land of Sequassen, the chief sachem of the tribe of
Suckiage, planted their colony, the first offshoot of *western* emigrants from the settlements of the bay, on the
immediate banks of the Connecticut, and on both sides
of the same Little River, or Riveret, as it is written in
the early records. Pushed to the north and west by the
sturdy occupancy of the Dutch, and warned early, doubtless, by the inconvenience and disasters of the spring and
autumn freshets of the Great River, and especially of the
sudden and higher floods in the season of growing crops,
the settlement spread out upon the securer terrace above,
although those future governors of the colony, Hopkins,

Part of the Ox Pasture

Th. Gridley

John White

John Moody

A. Smith
sold to
J. Barnard

John Barnard

Gregory Wilterton.

Moodys to Ox Pasture and Wethersfield

Sam. Greenhill heirs

Giles Smith

To Wethersfield.

Sam. Cole

Thom. Judd.

Charter Oak

Andr. Bacon

Geo. Wyllys Esq.
bought sev. lots

Nath Ward

Deac. And. Warner

American Elm

Sam. Wakeman

Wm. Hills to Ox Pasture etc.
also South to Wethersfield

Road from Charter Oak
also South to Wethersfield

Wm Hills

Wm. Hills

To Wethersfield to Ox Pasture

Thom Hooker

John White

Thom Wells

John Webster

Wm. Whiting
Merchant

Wm. Gibbins
of great wealth and
learning

Edw. Hopkins Esq.
bought 3 house lots

Cut down by order of
City Authorities

To the South Meadow Padman Landing

Giles Smith

Mill to the South Meadow a Highway on the Bank of the River.

Highway on the Bank of the River

RR Land

Th Lord

Th Standley

Wm Goodwin
old

Sam. Stone
Teacher

Thom. Hooker
Pastor

John Haynes

S O U T H   M E A D O W

Landing

LITTLE
MEADOW

W
S — N
E

To the Dutch Landing

Dutch Burying
Ground

Line of the River in 1824.

Dutchmens Land.

Dutch Point

G R E A T   R I V E R

October 31, 1687, the Charter of Connecticut
was hid in the trunk of this Tree.
It was resumed May 9, 1689.

HARTFORD,
South Side
IN
1640.

CHARTER OAK.

Wyllys, Welles, and Webster, located their house-lots on
the western edge of the South Meadow, and immediately
below the spot on which we are supposed to stand.

This portion of the South Meadow—commencing almost at the hearthstone of Governor Hopkins's house, and
nearly opposite the house-lot of Rev. Thomas Hooker,
and including the burial-grounds of aboriginal Indians
and early Dutch occupants, and all that was left of the
site and stockade of the House of Hope, and hundreds
of acres of land immediately adjoining, which for two
hundred years had been only partially available for agricultural purposes—the genius and enterprise of SAMUEL
COLT rescued from the capricious and devastating dominion of the floods of Connecticut River, and made part and
parcel of the residences, warehouses, and workshops of
the city of Hartford—at least as safe, for all time to
come, as a large part of the kingdom of Holland. By
constructing a dyke, seventy feet broad at the base, and
with a top surface as broad as the average of the streets
of the city, along the southern bank of Little River to its
junction with Connecticut River, and thence south on the
western bank of the same river for a half-mile, more or
less, and thence west to the terrace or upland before
spoken of, and higher in its whole extent of two miles
than the highest flood ever reached, Colonel Colt practically demonstrated the proposition, which an eloquent and
sagacious divine of the city, Rev. Horace Bushnell, D. D.,
about the same time announced as the theme of a discourse, "Prosperity—Our Duty"—that it was "the duty
of Hartford," at least of that portion of the city built
by him, "to prosper."

Erecting within the sheltering banks of the Dyke an

armory for the construction of the revolver which he
had invented and perfected, in the incredible short period
of time from the first of May, 1855, when the first machine was set to work, to January 10, 1862, the day of
his death, he not only achieved for himself and his family a colossal fortune (as distributed, of over five millions
of dollars) and a princely homestead, but gave profitable
employment to thousands of workmen, added largely to
the population and wealth of his native town, contributed
a new and effective weapon to the magazines of war, and
revolutionized the manufacture of fire-arms throughout
the world. As the most fitting tribute to his remarkable
genius, and an appropriate memorial of his public service,
we propose to describe briefly his great invention, the
Colt Revolver, and his great business enterprise, the Armory, with its ingenious mechanism for the construction
of his Fire-arm, together with the Dyke, and the crowning aim and result of all his labors, the Homestead,
within which he had garnered up his heart's best affections and aspirations. Before descending from the Hill
into the meadow, or *mear*, as our old Saxon tongue
would have designated this broad field or mead,—and
which, from its present and future association with the
Arm whose invention and construction have wrought the
wonderful transformation in its condition, we shall de-
nominate ARMSMEAR,—we will linger for a few moments
on objects and scenery which are honorably associated
with the subject of this memorial, and helped to inspire
the motives which led Colonel Colt to locate and build
his Armory and his Home here and not elsewhere.

No other spot is so near the well-springs or so full of
the traditions of our State history—so identified with

the race which, with their bravery, their rude culture,
their religious rites and horrible butcheries, have entirely
disappeared ; or with the authentic facts and veritable
representatives of that braver and wiser race, who, under
the leadership of Hooker, Haynes, Stone, Wyllys, Hopkins, and their associates, laid here the foundations of a
Commonwealth in which they aimed " to maintain the liberty and purity of the Gospel," and "to be governed and
guided by such laws, rules, orders, and decrees as shall
be made, ordained, and declared by the General Court to
be appointed by the freemen of the Commonwealth." In
the meadows below, in the rivers which are in sight, all
through the woods which darkened the uplands in all
directions, were the Indians;—within hail were their
fisheries, their cornfields, their hunting-grounds, their
burial-places, and their rude wigwams.

> " Dark as the frost-nipped leaves that strewed the ground.
> The Indian hunter here his shelter found ;
> Here cut his bow, and shaped his arrow true.
> Here built his wigwam, and his bark canoe,—
> Speared the quick salmon leaping up the fall,
> And slew the deer without the rifle ball.
> Here his young squaw her cradling-tree would choose.
> Singing her chant to hush her swart pappoose;
> Here stain her quills, and string her trinkets rude.
> And weave her warrior's wampum in the wood."

The tradition, which connects the tribe that sojourned
in the South Meadow with the old Oak which afterwards became historical, by a different tie than usually
binds the memory of the Indian with any object in
New England, should not be omitted here, as it has
passed out of the traditions and inspirations of poetry
into the keeping of history. Hollister, in his History of
Connecticut, adopting the words and authority of His-

toricus (Hon. I. W. Stuart), in his Life of Governor
Wyllys, who follows the first written version of this tradition by Mrs. Sigourney, who with true poetic instinct
had seized on this incident and preserved it in her " Intercession of the Indians for the Charter Oak," says:—
" We are told, while the manager of this property, Mr.
Gibbons, was preparing the house-lot assigned to Mr.
Wyllys for a dwelling and occupancy, he was waited
upon by a deputation of Indians from the South Meadow,
who came up to remonstrate against the cutting down
of a venerable oak that stood upon the side of the
mound now consecrated to freedom. With the true
eloquence of nature, the brown sons of the forest
pleaded in behalf of the immemorial tree. 'It has been
the guide of our ancestors for centuries,' said they, ' as
to the time of planting our corn. When the leaves are
of the size of a mouse's ear, then is the time to put the
seed in the ground.' At their solicitation the tree was
permitted to stand, and continued to indicate the time
when the earth was ready to receive the seed-corn; a
vast legendary tree, that must have begun to show signs
of decay a hundred years before that day, in the cavity
at its base, that was gradually enlarging as one generation
after another of red men passed from beneath its shadow."
The portion of Mrs. Sigourney's version of the traditional
intercession is as follows:—

> " Oh, not upon that mossy trunk
> Let the dire axe descend,
> Nor wreck its canopy of shade,
> So long the red man's friend-
> Nor to the cold, unpitying winds
> Those bannered branches give :
> Smite down the forest, if ye will,
> But let its monarch live!

" For far away, in olden time,
   When here the red deer flew,
 And with its branching antlers swept,
     In showers, the morning dew—
   Up like a solemn seer it rose,
     By hoary years unbent,
 Marking the sunshine, and the frost,
   Which the Great Spirit sent."

We can very readily believe that the Indians of this region had a feeling of veneration for this majestic oak, which even then was old,—for in this they only resembled their brethren in other parts of the land, whose traditions are connected with some majestic representations of the primeval forest; or the Druids of ancient Britain, who clothed the oaks of Berroc and Mona with a sacred character, and performed spells and sacrifices in the twilight depths of their congregated shades; or the still older Pelasgians, the aborigines of a portion of that country which was afterwards "mother of arts," who found the oracles of Zeus in the murmurs of the wind through the oaks of Dodona. Among every people, as soon as the poet appears to utter in winged words and natural symbols the highest and deepest instincts and aspirations of the human soul, the oak is seized upon as the type of manly strength, heroic resistance, venerable age, and earthly sovereignty. These characteristic features are admirably embodied in Virgil's description of the oak, in the *Georgics*, which he calls

"Jove's own tree,
 That holds the woods in awful sovereignty,
 Requires a depth of lodging in the ground,
 And next the lower skies a bed profound;
 High as his topmost boughs to heaven ascend,
 So low his roots to hell's dominions tend.
 Therefore nor winds nor winter's rage o'erthrows
 His bulky body, but unmoved he grows.

"For length of ages lasts his happy reign,
And lives of mortal men contend in vain.
Full in the midst of his own strength he stands,
Stretching his brawny arms and leafy hands;
His shade protects the plain, his head the hills commands."

Although we are prepared to believe in the rude Indians' veneration for our old oak, and even in their making an agricultural oracle of its young leaves, we doubt if their intercession was necessary to the salvation of a tree so old and majestic as this, with either Mr. Wyllys, or Mr. Gibbons, his manager, who had been brought up under the ancestral oaks of Fenny Compton, at Knapton, in the county of Warwick, England, and who probably selected his "house-lot" because of this and other noble forest-trees on its northern slope, as well as on the capabilities of the terrace above for "a garden of pleasaunce." Their intercession, backed by the tomahawk and scalping-knife, would have been better addressed to a manager and proprietor like some Street Commissioners, or Road Committee of the Common Council of a later day, who have deliberately cut down and wrenched up by the roots some of the noblest elms which ever graced the earth, to gain a few inches nearer approach to a dead level in a sidewalk, or to take out a graceful curve in a public thoroughfare, and bring all the intersections of streets to right angles. But whatever may have been the fact, or the manner or object of the intercession of the Indians, the Oak was spared.

"Yes, old memorial of a darkened age,
Thou lived to flourish in a brighter day,
And seemed to smile, that pure and votive vows
Were breathed where superstition reigned; thy trunk
Its glad green garland wore, though in decay,
And years hung heavy on thy time-stained boughs."

Than a tree a grander child earth bears not.
What are the boasted palaces of man,
Imperial city, or triumphal arch,
To forests of immeasurable extent,
Which time confirms, which centuries waste not?
Oaks gather strength for ages; and when at last
They wane, so beauteous in decrepitude,—
So grand in weakness—e'en in their decay
So venerable,—'t were sacrilege t' escape
The consecrating touch of time. Time watched
The blossom on the parent bough: Time saw
The acorn loosen from the spray: Time passed
While, springing from its swaddling shell, yon oak,

The cloud-crowned monarch of our woods, by thorns
Environed, 'scaped the raven's bill, and sprang
A royal hero from his nurse's arms.
Time gave it seasons, and Time gave it years;
Ages bestowed, and centuries grudged not.
Time knew the sapling, when gay summer's breath
Shook to the roots the infant oak, which after
Tempests moved not. Time hollowed in its trunk
A tomb for centuries; and buried there
The epochs of the rise and fall of states,
The fading generations of the world,
The memory of men.

No one spot in Connecticut is more widely known, or has been more visited by the leading spirits of the town and State, as well as by strangers and citizens from other parts of the country—drawn here by the historic and intrinsic attractions of the ground. For two hundred years it was the residence of one of our most cultivated and honored families. The original proprietor came here from an old ancestral hall and estate in the county of Warwick, in England, and brought with him hospitable habits, and a taste for gardening and agriculture; and, at the same time, means enough to continue his family habits and indulge his tastes from the start. He was the first to import and ingraft choice garden fruits, and to set out an orchard of apples; and the flower-garden and shrubbery and miniature lake of Wyllys Hill were for a long time the admiration of the Colony. In his immediate neighborhood were the homes of Hopkins and Wells and Webster, of Haynes and Stone and Hooker, and other kindred spirits who had come over together, or for a common object, and who would naturally seek each other's society, in their isolation from other settlements, and from their distant fatherland. His relations, as well as those of his descendants, to the government of the Colony, brought him and them into frequent and intimate intercourse with the prominent men in public office and the ministry from other parts of the Commonwealth down to the death of the last Secretary, in 1823. The head of the family, George Wyllys, was Governor, as well as Deputy Governor and Assistant; the second, Samuel, a graduate of Harvard, was thirty-six years Assistant, and four years one of the Commissioners of the United Colonies; the third, Hezekiah, was Secretary of

4

State for twenty-three years; the fourth, George, who
graduated at Yale, in 1729, was in the same office sixty-
one years; and the fifth, Samuel, who graduated at
Yale, in 1758, held the same office thirteen years. "It
is believed," remarks Mr. Stuart, in his *Lives of the
Early Governors of Connecticut,* "that this instance of
the perpetuation of high office in the same family for
so long a series of years, is without a parallel in this
country." It is a creditable instance of the steady habits of Connecticut, when, as in this case, those habits
were set in the right direction. Our object in referring
to it here is only to show the almost necessary connection which the residence of these successive Secretaries
of State must have had with those who were associated
with them in the administration of the government,
and with the people of Connecticut, for nearly two hundred
years.

We have before us—in a letter addressed to the President of the Connecticut Historical Society by Mrs. Anstes
Lee, in 1855, then living at Wickford, R. I., in the full
possession of her memory, and who died only two years
ago (July 10, 1864), within a few months of being one
hundred years old, in consequence of a fall, and not from
age—a pleasing picture of the Wyllys family, and of the
interest attached to this spot. In May, 1791, Mrs. Lee
made the journey to Hartford, on horseback, in company
with her brother, Daniel Updike. We shall, we trust, be
pardoned if we preface her account of her visit to the
Wyllys family, and of her meeting there the Rev. Dr.
Stiles, President of Yale College, and Colonel Ingersoll,
the State's Attorney for the county of New Haven, and
other gentlemen, and of their all going out to stand

under that famous Charter Oak; and while there listening
to Dr. Stiles's narrative of the seizure and preservation of
the Charter, by a few extracts descriptive of manners and
times now gone by.

"In conformity to previous arrangement, on the Monday of your election week, 1791, my elder brother,
Daniel Updike (who lately died at East Greenwich,
in June, 1842, at the advanced age of 81 years), and
myself started on a visit to Connecticut. We left our
father's house, the residence of the late Lodowick Updike,
near Wickford, on horseback; carriages at that time being rarely used, as the roads were so bad that it was
impracticable to travel on them with comfort or safety.
I was mounted on a fine Narraganset pacer, of easy carriage and great fleetness; she was the last of the pure
blood and genuine gait that I have seen. We arrived
at Plainfield village on the same afternoon, and lodged
at Judge Lightfoot's. The Judge had been a resident of
Newport for many years before his removal to Plainfield. He was an intimate friend of my father, and had
visited our mansion in the days of my grandfather,
Daniel Updike, who was for twenty-seven years the Colony Attorney-General of Rhode Island. Judge Lightfoot
was an Englishman, educated at Oxford, studied law at
the Inner Temple, and was subsequently appointed a
Judge of Vice-Admiralty for one of the Southern Colonies. We spent a very social and pleasant night with
our friend, who appeared to be much gratified with a
visit from his Rhode Island acquaintances.

"On Tuesday morning, after a cordial shaking of hands
and with his best wishes for a pleasant journey, we left
for Hartford. We took the usual route through Wind-

ham and lodged that night at a public house in Bolton, kept by one Mr. White, twelve miles short of Hartford. We rose early on Wednesday, and arrived at Hartford and put up at Bull's Tavern (sign of the bunch of gilded grapes), and took breakfast on *bloated* salmon. I particularly recollect about the salmon, as it was the fashion in early times for parties of gentlemen of Rhode Island to make a special visit to Hartford almost yearly to luxuriate on this rare and delicate fish, which at that period were there in great abundance, and rarely any in the Narraganset rivers. While we were at breakfast, Mr. Ralph Pomeroy came to take us to his house, which stood on a street near where the Episcopal church then stood. His wife was the widow of William Gardiner, brother to my mother. Her maiden name was Eunice Belden; he was killed by the explosion of the powder-house at Hartford, on celebrating the repeal of the Stamp Act. He left a son by the name of James Gardiner, who died some forty years ago at Hartford. Mr. Pomeroy had been commissary in the Revolutionary War, and had practised law with repute, as I understood. He had been frequently at my father's in Rhode Island.

"This being the day previous to the general election, the city became quite thronged with people from all parts of the State. In the afternoon, Governor Wolcott was expected to arrive. This event seemed to awaken great interest and appearance of parade. To witness the display, Mr. Pomeroy took us to the house of General Wyllys, which stood nearly opposite to the State House. General Wyllys was the son of old Colonel Wyllys. He appeared to be a fine gentleman, aged about forty; his wife was Elizabeth Belden, sister of Mrs. Pomeroy.

"A troop of horse and a great number of citizens on horseback constituted the cavalcade to escort his Excellency into the city. The company of horse made an imposing appearance. The riders were dressed in caps, with a brass plate, and feathers in them, short jackets, or coats, short-clothes and high gaiters. I think the color was deep blue faced with red. The horses were very fine, and Mr. Pomeroy said they were of two hundred dollars value each, which was a great price at that time. After tea, say an hour before sunset, it was announced that the Governor and procession were entering the city; all thronged the windows to view it. The Governor came in at the head of the military on a single horse, dressed in a full suit of black, and then followed the cavalry and the citizens on horseback, two and two abreast. When he arrived in front of the State House he alighted, ascended, and stood on the spacious front step. The military passed and saluted the Governor by a discharge of their pistols over his head. After the salute, the Governor walked to a public house near.

"The next day (Thursday) was the general election. The General Assembly organized and proceeded to the meeting-house to hear the election sermon. It was the longest procession I had ever seen. It was headed by the military; then followed the Sheriff with his sword, the Governor, Senate, and Members of the House of Representatives, two and two; and then singly walked President Stiles, dressed in a full black gown, cocked hat, and full-bottomed white wig. Then came the clergy, two and two. I should think there were two hundred ministers, dressed in black, and after them walked the citizens. Such an imposing procession I had

never seen. I did not attend church, on account of the
crowd. The Legislature convened again after service and
refreshments, which were furnished at State expense, as I
was informed. It was carried into the State House on
trays.

"Friday afternoon, the day after election, upon invitation, we spent with Colonel Hezekiah Wyllys, who lived
at the Charter Oak Place. It was an ancient-looking
mansion, that stood on a square by itself. The view
from it was splendid. It overlooked the Connecticut
River, and you could see its serpentine course for a
great distance, and you could also overlook the city.
The great Charter Oak stood right before it. We were
shown the grounds and the garden, which were beautiful
and tastefully arranged, with many flowers in full bloom.
Colonel Wyllys was uncle to Mrs. Pomeroy. Her sister
married a Mr. Belden, and General Wyllys married Miss
Elizabeth Belden, his cousin. I should think the Colonel was then over seventy, thin and spare, with baize
round his feet.

"President Stiles, Colonel Ingersoll, the Attorney-General of Connecticut, and several other gentlemen, took
tea at Colonel Wyllys on the same afternoon. President
Stiles and my brother conversed nearly an hour about
Rhode Island. He had been settled over a Congregational church in Newport many years, and was much at
Colonel Willet's in Narraganset. Mr. Willet was uncle
to my mother. I remember seeing Dr. Stiles once at
St. Paul's Church, Narraganset, where Dr. Smith (afterward Superintendent of Cheshire Academy) was rector.
My father was introduced to him after service. We all
went out after tea to see the Charter Oak, and stood

under it; Colonel Wyllys was too infirm to accompany us. I felt anxious to stand under the celebrated old tree, where the old Colony Charter was hid by the ancestor of the present occupant. President Stiles gave us (we standing round him) a minute and detailed account of all the circumstances of its seizure and concealment; his manner was eloquent, and the narrative was precise and particular, and it made a deep impression on me. It is fresh in my recollection now, although a half century has passed away since I heard him. I well recollect his sharp face, spare person, and precise manner.

"The mansion of Colonel Wyllys I admired, and the manners of the Colonel's family combined urbanity with dignity. The room where we sat was spacious, and there was a greater display of silver than I had before seen. There was a large mahogany table in the parlor, and under it stood a finely wrought silver chafing-dish, and a silver tea-kettle stood on it, that coal might be placed in the chafing-dish, and by that means the water for the tea might be kept hot; there was also a large silver tea-urn. On the table stood a large silver waiter, and a large silver tea-pot, silver sugar-dish, and silver cream-pot. This was surrounded by a richly ornamented set of China service; in unison with that were elegant carpets, chairs, and mirrors. It was impressive evidence of an ancient family of wealth."

In the absence of any printed or written record of the results of Dr. Stiles's well-known habits of investigation into local traditions, we give a letter of Hon. Isaac W. Stuart on the historical teachings of the old Oak:

LETTER OF THE HON. I. W. STUART TO MASTER SAMUEL JARVIS COLT,
ON PRESENTATION OF CHARTER OAK CRADLE.

HARTFORD, *April* 25, 1857.

MASTER SAM. JARVIS COLT:—You are a tiny infant now, Sammy, just bursting into life, and cannot read or understand what I write you at the present time. But you will grow, and be able to read and understand it by and by, and then you will find out that a friend of your father and mother and to you, sent for your use this day a beautiful present—a cradle for you to rock in when you were a baby, and a cradle made from the wood of a very famous tree called the *Charter Oak*.

This tree is famous, Sammy, because a long, long time ago, when the State in which you were born was as young almost as you are now, a very bad man, named Edmond Andros, came with a troop of soldiers to the town in which you first saw the light, and tried to take away from the good people there, a long parchment roll which was called the Charter of Connecticut.

Now, this Charter made the people of whom I speak a very free people. It gave them *Liberty,* and *Liberty,* you will live, I hope, to learn, Sammy, is a very precious thing, and ought to be defended at the cost of all the money in the world—yes, and at the cost, too, sometimes, even of human life.

But the people of whom I speak were few in number and weak, and no match at all in power for that man who came to take away their Charter, for this bad man was backed by a great tyrant king, who lived far away across the rolling sea, and who had armies and navies big enough to crush the little settlement in which these people lived, forever. So Edmond Andros thought he was sure of getting the Charter, because these people were weak, and he and his master were strong; and he went, therefore, with a file of soldiers into a large room, called the Court Chamber, where the Governor and Deputy Governor, and the great men of these people were assembled—there he went to take the Charter. But just as he moved forward to a large round table on which the Charter lay, and was stretching out his hand to snatch it, the candles in the chamber were all suddenly blown out, as quick as thought, and a very brave, patriotic man, named Captain Joseph Wadsworth, who loved the people and the liberty of which I speak, seized the Charter himself, in the midst of the darkness, and running with it as swift as a deer, just as you will run, I hope, by and by, Sammy, he hid it in the hollow of a beautiful great oak; and when the candles in the chamber where Edmond Andros was, were lighted up again, lo, Sammy, *the Charter was gone!* and Andros couldn't find it anywhere, and never did, and the people of whom I speak were happy again, and rejoiced because their liberty was saved by means of this beautiful great oak; and they loved this *Oak* forever after.

Now, Sammy, your cradle is made out of the wood of this beautiful great oak, and it should teach you always to remember that hero who hid the Charter so well, and make you follow his example in defending your country whenever it is in danger. *Die for your native land*, Sammy, *rather than let anybody hurt it!*

This cradle too should teach you to be wise, and

5

virtuous, and honorable, and industrious, just as those
good men were who lived when the Oak was made so
famous. They thought it was best always to do right
—so always think yourself. They adored liberty—so do
you always adore liberty. They worshipped justice—so
do you always worship justice. They made truth their
idol—so do you always make truth your idol. They
worked out their own prosperity—in other words, these
good men of the olden time "paddled their own canoes"
—so always do yourself; and mark it, Sammy, that I
caused your cradle to be made in the shape of a beautiful *canoe*, in order to remind you of this sterling duty.
Your father thought a great deal of this duty, and he
acted it out, and once told a great committee of the
lofty British Parliament that "*he paddled his own canoe.*"
Sammy, remember it!

But first, and last, and best of all, my little friend,
those good men of whom I speak, who lived in the
days when that oak of which your cradle is made first
became famous, loved God, that great and good being
above us, far away up in the golden skies, who takes
care of us all, and wants us all to love him, and if we
do so, will make us all happy. Sammy, love God, love
your father and mother, love your Aunt Hettie, who is
all the time now finding pretty dimples in your cheeks;
love all your kindred, love all mankind. "Be virtuous,"
Sammy, as your father often says, and then you will
certainly be happy, and go some day and see God.

From the old proprietor of the Oak,
And your true friend,
I. W. STUART.

THE cradle itself is canoe-shaped, the base being a natural limb of the oak—from the bottom a beautifully carved acorn depends, to which are attached two cords, used in swinging it. The sides are elaborately carved in open work, representing branches, leaves, and acorns of the oak; in the centre, on one side, is Col. Colt's coat of arms, and on the other a tablet for an inscription. The posts supporting the cradle are finely carved from a solid piece of the oak, imitating the intertwining branches, the whole surmounted by two colts rampant, facing each other, which produce a fine effect. The platform on which it stands is veneered with the genuine bark of the Charter Oak—in the centre of this base is a noble knot of the oak, upon whose polished lower surface, in Mr. Stuart's handwriting, are the following verses from Mrs. Sigourney's poem on its fall:—

"That brave old Oak
Stood forth, a friend indeed,
And spread its Ægis o'er our sires
In their extremest need;
And in its sacred breast
Their germ of Freedom bore,
And hid their life-blood in its veins
Until the blast was o'er.

"The fair ones of our Vale
O'er its fallen Guardian sigh—
And Elders, with prophetic thought,
Dark auguries descry:
Patriots and Sages deign
O'er the loved wreck to bend,
And in the funeral of the Oak
Lament their country's friend."

Various parts of the cradle have been since inlaid with amethyst, topaz, and other precious stones, purchased by Col. Colt at the Asiatic fair at Novgorod.

This valuable present, most artistically designed and skilfully wrought, out of material as precious as the cedar of Lebanon or the wood of Shittim, which the above letter, accompanying the same, so beautifully describes, and whose patriotic lessons, drawn from its historical associations, it so impressively inculcates for the instruction of every child of Connecticut, was doubtless a graceful recognition, on the part of Mr. Stuart, of the deep interest which the father of young Sammy Jarvis Colt had always manifested in Charter Oak when standing, and in the preservation of its shattered trunk when compelled to bow in its waning strength to the violence of the tempest. He had known the old tree and the Hill where it stood long before Mr. Stuart came to reside in Hartford. As a boy, in common with other boys of his age, he had doubtless gathered its acorns with their cups for his sisters; he had coasted in winter down the long slope of the hill into the meadow road, and skated on the "Wyllys pond," as other boys were permitted to do by the last male representative of the Wyllys family. At that time, the old oak did not exhibit such signals of decay and dilapidation as it put on some thirty or forty years later—owing in part to the neglect which the dilapidated fortunes of the family and want of male guardianship for a period subjected it to, and partly to the injury done by a severe sleet and ice storm to its twigs and outer branches.

A writer in Harpers' Magazine for May, 1862, in an interesting article on "American Historical Trees," in a brief description accompanying a sketch of the oak as it appeared in September, 1848, "after a gale of nearly thirty hours' duration had stripped the oak of nearly all its leaves, covering the ground beneath with foliage and acorns," cites a letter written by a lady sixty years before, thus:—"Age seems to have curtailed its branches, yet it is not exceeded in the height of its coloring, or richness of its foliage. The cavity which was the asylum of our Charter was near the roots, and large enough to admit a child. Within the space of eight years that cavity has closed, as if it had fulfilled the divine purpose for which the tree had been reared." The writer was mistaken as to the position of the cavity which received the Charter, and although the orifice to which she referred was closed, the corroding processes of decay were at work at the heart of the gnarled and unwedgeable trunk. When Col. Colt returned to Hartford to commence the manufacture of his Revolver, it was fast becoming the veritable reality of that fine picture of Spencer:—

> " A huge oak, dry and dead,
> Still clad with reliques of its trophies old,
>   Lifting to heaven its aged, hoary head,
> Whose foot on earth has got but feeble hold;
>   And, half disbowelled, stands above the ground,
> With wreathed roots, and naked arms,
>   And trunk all rotten and unsound."

But it was still the old historic oak of his boyhood, venerable in its centuries of age and majestic in its bulk —measuring thirty-six feet in circumference at the ground, and twenty-five feet measured at the smallest part above. Its boughs, although greatly thinned, and shorn of their extreme length, still had the characteristic features of its best days, some stretching out almost horizontally to a vast distance, while others, crooked and bent, put out to the right and left, upwards and down-

wards, abruptly or with gentle sweep, as they did centuries ago, when its tender leaves were so carefully watched
by the Indian, for the season to drop his seed into the
earth. It still stood—Spencer's "sole king of forests all,"
or Virgil's "tree sacred to Jove." Its story had passed
into the keeping of history, and become the inspiration
of the poet and the orator; it was told at every fireside,
and studied and recited in every school of the land.

We confess to a fellow-feeling with Col. Colt in his
admiration and passionate fondness for old trees, be they
oak, or beech, or elm, or any other species, so be they
are strong beyond the power of words to express, in their
spread and depth of roots, and breadth and sturdiness of
trunk, or majestic in their towering height, or graceful in
their sweep of boughs, or rich in their wealth of foliage,
and venerable with that sanctity which age alone can
give, for, with trees at least,

> "Time consecrates, and what is gray with age
> Becomes religion."

Standing hundreds of times beneath this old tree, we
had no need of any other reminder of the Great Architect whose hand alone could rear through the centuries
its mighty fabric, and weave year after year its verdant
roof, but with Bryant could exclaim—

> "Thou hast not left
> Thyself without a witness, in these shades,
> Of thy perfections. Grandeur, strength, and grace
> Are here to speak of thee. This mighty oak—
> By whose immovable stem I stand, and seem
> Almost annihilated—not a prince
> In all that proud old world beyond the deep,
> E'er wore his crown as loftily as he
> Wears his green coronal of leaves with which
> Thy hand has graced him."

But time and tempest did their worst on the old
Oak; and the fell spirit of speculation, more reckless of
all objects of sentiment, veneration, and patriotism, which
do not minister to its hungry greed of gain, than the corroding tooth of time or the sharp violence of the tempest,
not only swept from the hill every vestige of garden
and lake, of mansion and tree, but even wrenched out
of the soil by machinery, the very roots whose slowly
decaying fibres should have been left to consecrate forever
the soil, which, for untold centuries, had nourished its
majestic trunk and more than kingly coronet of leaves.
Not content with this, but when a slight deviation in the
line of the road would have rendered it unnecessary, the
same spirit, to gain a few more feet of building-ground,
worth a certain number of dollars in the market, carried
the wheels of daily travel and traffic directly over the
spot where that majestic trunk stood. Instead of a noble
monument, with a suitable inscription to perpetuate the
heroic love of liberty which that tree symbolized, a slab
hardly sufficient to mark the grave of an infant was inserted in the road-bed. All this was done in open day,
with a cloud of witnesses looking sadly on; and so well
satisfied were these "architects of ruin" with their work,
that they perpetuated both the act and themselves in a
photographic picture of the scene! Fit companion of that
other scene in our local history, the destruction of scores
of noble elms which protected our sidewalks from the
blazing sun of summer, and imparted a picturesque effect
to the monotonous architecture of our streets. The only
voice of remonstrance and indignation from the public,
whose patriotic memories were thus outraged, came trum-
pet-tongued from the heart of Mr. George H. Clark :

THE OAK.

" Yes, blot the last sad vestige out—
  Burn all the useless wood;
Root up the stump, that none may know
  Where the dead monarch stood.
Let traffic's inauspicious din
  There run its daily round,
And break the solemn memories
  Of that once holy ground.

" The hallowed spot your fathers long
  Have kept with jealous care,
That worshippers from many lands
  Might pay their homage there;
You spurn the loved memento now,
  Forget the tyrant's yoke,
And lend oblivion aid to gorge
  Our cherished Charter Oak.

" 'Tis well, when all our household gods
  For paltry gain are sold,
That e'en their altars should be razed
  And sacrificed to gold.
Then tear the strong tenacious roots
  With vandal hands away,
And pour within that ancient crypt
  The garish light of day.

" Let crowds unconscious tread the soil
  By Wadsworth sanctified;
Let Mammon bring, to crown the hill,
  His retinue of pride;
Destroy the patriot pilgrim's shrine--
  His idols overthrow,
Till o'er the ruin grimly stalks
  The ghost of long ago.

" So may the muse of coming time
  Indignant speak of them,
Who Freedom's brightest jewel rent
  From her proud diadem;
And lash with contemptuous scorn
  The men who gave the stroke,
That desecrates the place where stood
  Our brave old Charter Oak!"

This Hill, with all its commanding outlook of near and distant scenery, so accessible to the main thoroughfare of the city, and so directly associated with the history of the State, it was the patriotic desire of Col. Colt to preserve, in its natural slopes and surfaces, and with as many of its historic memorials as possible, for public use and enjoyment, as a promenade and as the site of a new State House. For these objects he made strenuous efforts, and expressed his willingness to be taxed, in common with other property-holders, to any extent the city or State authorities should authorize, and to contribute largely from his own resources besides. But his efforts were not adequately seconded by others, and although a resolution authorizing the building of a new State House on Charter Oak Hill passed one branch of the Legislature (the Senate) in 1857, the movement failed. A young tree raised by a citizen from an acorn of that historic oak he welcomed as a most acceptable present, and planted before his favorite window—

> "Glad that some seedling gem,
> Worthy such noble stem,
> Honored and blessed in his garden might grow."

How thrillingly would he gaze on it to-day—the green of all its outer leaves tinged with golden halos, tokens that his wishes for it are fast fulfilling.

> "Heaven send it happy dew!
> Earth lend it sap anew!
> Gayly to bourgeon and broadly to grow."

The spirit in which Col. Colt selected and improved these meadows, as the site of his great Armory, and as part of his homestead, is strikingly set forth by Mr. Stuart, in his discourse at the dedication of Charter Oak Hall:—

6

"Through a truly commendable judgment of his own, at the very start of his great enterprise upon these meadows, he decided that all its principal structures, its avenues, its streets, its docks, its areas, if any, that might be reserved and embellished for public purposes, should, as from time to time the new improvements might appear, receive commemorative names—names that are replete with significance—that should remind of the past— that should interweave with the present—and carry the imagination forward, with gladsome anticipations, to the future upon this the seat of his proprietorship and of his laudable pride. And the result has been thus far—

"*First*, that the aboriginal proprietors of these meadows and this town, the Indians of the tribe of Suckiage, are commemorated through the names of five of those the representatives of their race who deeded Hartford to its first English settlers. *Sequassen* Street preserves the memory of their proud, valiant, persevering, wary, yet faithful head sachem; *Wawarme* Avenue, that of his sister and only heir; *Masseek* Street, *Weehassat* Street, and *Curcombe* Street, that of three of his successors, subordinate sagamores, whose grant, in union with that of some others of the Indian blood royal, gave the whole beautiful area, extending from Wethersfield on the south to Windsor on the north, and from 'the Great River' on the east 'full six miles' into the wilderness on the west, to the founders of our town, the memorable emigrating band under Hooker and Haynes.

"*Second*, it has resulted from the plan under consideration, that the Dutch, who were the first of Europeans to ascend the Connecticut, and build, and possess, and plant upon these 'South Meadows,' are also commemorated.

Huyshope Avenue recalls at once the fort they constructed just by us, at the mouth of the Little River, or *Riveret*, as it is more beautifully denominated in our earliest records. *Vandyke* Avenue—that directly confronting the river—preserves the name (Gysbert Vandyke) of the original commander of this small, yet compact fort, from whose ramparts the Dutch flag floated in pride for about twenty years; and at the same time, by a happy double meaning in its last syllable—by an accidental yet felicitous paronomasia—designates in English that gigantic embankment around us now, which has shut out the waters of the Connecticut in their maddened freshet-time, and, like corresponding constructions in Holland against old Ocean's inroads, has 'to the stake a struggling area bound.' *Vredendale* Avenue and *Vredendale* Dock—or *Peacedale* Avenue and *Peacedale* Dock, as beautifully in our own vernacular this Dutch appellative signifies—notify us of the supreme governor, appointed by the 'High and Mighty States General' of Holland, over the fort upon this spot, Johannes De La Montagnè, a member of the council of the New Netherland, a doctor of medicine, and the owner of a blooming country-seat, designated by the soothing name of *Vredendale*, which lay within the circuit of the present Empire City of the Union—that city, be it marked, from whose port the yacht *Onrust*—the first built vessel of that '*restless*' metropolis, whose enterprising commerce now 'defies every wind, outrides every tempest, and invades every zone'—pushed the first voyage up our own Connecticut, and bore the great navigator Block to discover and map out the site of our own Hartford. This navigator, and his well-famed lieutenant, Hendricxsen, are also suitably commem-

orated in avenues respectively entitled *Van Block* and
Hendricxsen.

"*Third*, it has resulted from the plan under consideration, that our English progenitors, who purchased here of
the native inhabitants, and who supplanted the Dutch—
by virtue, as I think, of a just title, originating in the
right of prior discovery, or if not in this, then in the
right of legitimate conquest—are also, soon as certain new
streets, already laid out, shall come to be improved, to be
commemorated here through the names of their *Hooker*,
their *Haynes*, their *Hopkins*, their *Webster*, and others,
their choice and leading men—one of whom only, thus
far, is memorialized, and in the name of *Wyllys* Avenue.

" And *fourth*, it has resulted, that an ever-memorable
contest for liberty, in the infancy of Connecticut, between
these our English progenitors on the one hand, and their
parent country on the other—in which a monarch tree
figured as the deliverer of the oppressed—is also commemorated in the name of this the avenue upon which
we are now assembled, and at the head of which that
deliverer stands—the *Charter Oak* Avenue.

"So it happens—by virtue of the nomenclature thus
far applied—that the Indian, the Dutch, and the English
antecedents of this spot are all indicated; that a three-
fold history is signified, each part of which, by itself, is of
deeply interesting import, but which, compounded, forms
one of the most remarkable and thrilling pictures on
record of human experience in the colonization and settlement of this new world; and that the whole of this
great past, by means of another and a modern designation—the only one which in justice ever could be
thought of to mark that gigantic structure yonder and its

projector—is made to link in happily with the present. This modern designation is COLT'S ARMORY!

"Upon one of the panels, on the eastern wall of this apartment, has been painted, it will be observed, with great accuracy, a map of Hartford as it was in 1640, during the period of its first settlement; while upon another and adjacent panel, that portion of the South Meadows now occupied with the improvements of Col. Colt has also been delineated, as it was when purchased by him, with the names of the proprietors from whom he bought upon the respective lands which they possessed; and upon a third panel it is contemplated also to exhibit the whole territorial site in question as it now is, distributed into avenues, streets, and areas;—so that, most happily, the walls of this Hall will themselves be made to express, in an imposing and instructive form, lessons of history and of material improvement.

"This Hall, with special reference to the population now and hereafter to be gathered in this valley, is intended, in the first place, for a *reading-room*, where newspapers and periodicals of an instructive character are to be collected for quiet perusal in hours not devoted to mechanical labor. It is intended also as a room where, occasionally, as opportunity may direct, *lectures* may be delivered, experimental or otherwise, upon science and art, upon philosophy and morals, or upon any topic where the purpose shall be to communicate useful knowledge. It is to be used, too, as a room for *discussion* or *debate*— should any associations here, within the circuit of these South Meadows, be organized for such an end—and as a room besides for the display of such interesting and improving curiosities and pictures as the good judgment of

its projector may from time to time select and appropriate, out of the stores which his own abundant means
enable him readily to accumulate.

"This place also is to be used as a room where parties may assemble 'to trip the light fantastic toe'—where
Colt's capital may gather, so it pleases, 'its beauty and
its chivalry,' and all go 'merry as a marriage bell'
upon a Thanksgiving or a Christmas Eve—or on the
night of an eighth of January, or a twenty-second of
February, or of an election day, or of the ever-memorable Fourth of July, or upon occasion of any holiday
celebrations where the object shall be innocent enjoyment. *Fairs*, too, designed to answer some special phi-
lanthropic end, may here make their display. *Concerts*
may be given here—finely foretokened to-night by stir-
ring harmonies from that band, now present, which has
been originated and munificently endowed by Col. Colt
himself, for the worthy purpose of domesticating high
musical art within this armory neighborhood. And to
tones of a far graver character than those which issue
from the lips of song, or from the mouths of silver instruments, these walls, within the purpose of their owner,
may sometimes be allowed to echo—ay, even to the
tones of religious teaching, should some 'reverend cham-
pion' perchance, of 'meek and unaffected grace,' desire,
here upon the new dwellers of this valley, to

'Try each art, reprove each dull delay,
Allure to brighter worlds, and lead the way.'

"Thus much in explanation of the purposes for which
this Hall has been established. It remains now only, in
conformity with the request of the projector of this structure, that I should bestow upon it a name. To this

duty then I turn. And yonder, in near neighborhood to this our place of assemblage—in loving familiarity, for hoary centuries, with the aspects of this valley, alike in sunshine and in storm—alike when it was dotted with the wigwams of the red man, and when our own ancestral pale faces first came to claim it—and now that it promises soon, its agricultural fast yielding to other uses, to become crowded with workshops and habitations—yonder, in solitary grandeur, almost overshadowing us from its own hill-top, and overlooking the majestic Connecticut, and the new and wonderful creations which the might of manufacturing art has here but as yesterday evolved—yonder, in the glory of a patriotic history that is unmatched by aught else of its own nature upon earth, stands that monarch tree, the CHARTER OAK! It has already often vouchsafed its opulent, expressive name. Our State is gloriously known in history, and in common parlance is proclaimed, as the *Charter Oak State*. The river and the sea have borrowed the name for the steamer and the ship. Organized associations—the bank, the insurance company, and the lodge—employ it. It circulates, beautifully impressed, on their bills, their life policies, and on their badges. It glitters, in its gilded symbol, on the accoutrements of the military company. It figures on the hotel, the store, the refectory, the saloon, and the market-place. It has long distinguished a leading street of our city—that upon which, continued in avenue amplitude down to this the great river of New England, we are assembled to-night. Yet, though often used, the good name never degenerates into triteness. It has, fortunately, resources of meaning that no consumption can waste.

"Under the authority then to me committed by the lawful proprietor of this edifice—the first-born structure, as, felicitously, it happens to be, upon this new-made avenue—and by virtue, too, of my own particular warrant as the lawful proprietor of the Oak itself—in view of the truly useful, liberal, and gladdening purposes which this Hall is designed to answer—considering also gravely the fact, vouched for by its sponsor here, that the spirit which is to reign within its walls assents immovably unto all the articles of the Charter Oak faith—I do therefore hereby name, proclaim, and publish this spacious apartment to the world as the CHARTER OAK HALL!"

But Col. Colt's aspirations were not limited to achieving for himself a name and a fortune in the place where he was born, and on a site associated with his boyhood's dreams of wealth and patriotic devotion; he desired to elevate the whole laboring class to a higher plane of intelligence, enjoyment, and effort, and to make their homes healthy, happy, and hopeful, for themselves and their children.

"It was the dream of his boyhood," remarked Col. Deming, then Mayor of the city, at the dedication of Charter Oak Hall, "a dream which, under the most extraordinary difficulties, discouragements, and reverses, sustained and cheered him—that if Providence should ever smile upon his industry and energy, he would here, upon this very spot (in the South Meadow, near the junction of Little and Great Rivers), rear an establishment which should not only be an honor to his native town, but a light, a landmark in the weary and disheartening pilgrimage of mechanical genius. We have before us a splendid realization of the poor boy's dream."

II. ARMSMEAR.

ARMSMEAR.

A PARTY, gay, grave, and venerable, as told in the letter of Mrs. Lee, which we were first to print, in the preceding chapter, stood under Charter Oak some seventy years ago. While then and there gazing around, they heard President Stiles describe how that tree had become ever memorable. That party, standing to-day on Wyllys Hill, with the same sky above, the same wooded hills in the distance, the same glimpses of river below, would behold around and below metamorphoses no less strange than any that are chronicled in Ovid.

Their horses' hoofs would tread on the site of the Charter Oak, the local habitation of which is marked only by a loose stone no larger than a cheese; but they might hear of that tree as re-living in a thrifty son, on the Colt lawn, as well as itself going abroad into all lands, fashioned into canes, chairs, crosses, and other curiosities, larger and smaller.

In the successor of the Wyllys mansion, should they be thirsty, they would discover no well—yet water, both hot and cold, would gush forth for them, even in the uppermost chambers, at the turning of a faucet. Light and heat would sally forth as readily and as promptly for their aid and comfort.

No longer need they wait in suspense for half a day, doubting when the Governor would arrive, but the telegraph would certify them of the very minute at which they would welcome him, and that borne to them swift as an eagle, and as it were in a runaway village. Nor would a journey from Rhode Island cost him, as theirs did, three days' riding, but within that time he could reach them from a point far beyond the farthest limits, south and west, of the Union as they saw it. Yes, on the fourteenth of August, 1866, they would have read, at the tea-table in Hartford, news which left London at noon of that self-same day, namely, that Prussia refused to surrender her Rhenish provinces to France.

None of these marvels, however, would amaze them more than the outlook over the South Meadow. The steam-gong and the volleys of the proving-house might startle them, as if war-whoops had wakened anew, and that intensified by two centuries of slumber. But they would soon learn that Indians had been supplanted by engines, scalping-knives by sword-bayonets, bows and tomahawks by revolvers and rifles.

Some trees which, as saplings on the hill, on the banks of Little River, or in the meadow, had witnessed their visit in the last century, remain in sturdy strength to welcome them back. Nothing of this sort, perhaps, would charm them so much as the Ledyard Elm, in front of the old Seymour mansion, which in its youth might have reminded them of the solitary voyage of the great traveller, in a canoe of his own construction, down the Connecticut from Hanover, in 1771, and the memory of which it was planted to commemorate — as, to its latest age, it must remind posterity of his, the great traveller's, tribute to woman — a eulogy more full of meaning than the ode of Schiller, because uttered out of a richer and wider experience. Some of `his words were: — "I never addressed myself in the language of decency and friendship to a woman, whether civilized or savage, without receiving a decent and friendly answer. With man it has often been otherwise. * * * * If hungry, dry, cold, wet, or sick, woman has ever been friendly to me, and uniformly so; and, to add to this virtue, so worthy of the appellation of benevolence, these actions have been performed in so free and so kind a manner, that, if I was dry, I drank the sweet draught, and, if hungry, ate the coarse morsel, with a double relish."

Below, instead of hay and cornfields, the long lines of the Arms-factory, and the dwellings of several hundred families, would rise beneath their gaze, all safe and secure, dry like Gideon's fleece while all was wet around, and that in a freshet, when the Great River, coming up like the swelling of Jordan, spreads itself broad as a lake on all the southern lowlands — not without often sweeping away their crops and fences in its current.

They would surely hasten down into the lowlands for exploring the causes of this phenomenon. In Potsdam they hear Teutonic speech, see its outlandish architecture, survey a veritable Dutch dyke, willows and all, a gigantic gray guardian girdling the meadow. They are more than half persuaded that, though Yankees and yearly overflows had "beat the Dutch and drove them out of town," yet that those indomitable dyke-diggers have returned to plague the victors, have retrieved their fortunes, and created a new world for themselves in that meadow where they first landed in New England. They halt at the triangular brick block, and are told of CHAR-TER OAK HALL, to which its upper stories are dedicated, — a structure reared for advancing the intellectual and æsthetic culture of mechanics, by lectures, concerts, a reading-room and debating-club, and at the same time their social enjoyments through festive assemblies. They enter the Armory, and are more than ever bewildered in wild-eyed wonder, as they walk among its machines of strength, dexterity, or speed. Deucalion could not have been more astonished at seeing all the stones he threw over his shoulder at once transformed to men. Then they look up to the Colt château, high-gleaming from afar.

To their Puritan simplicity, its massive towers, airy pinnacles, plate glass, and gilded finials, appear a fairy palace—

> "Such state as, somewhere in the East,
> Rose when Aladdin rubbed his wondrous lamp."

They would make haste to climb what would seem to them a "Delectable Mountain," — linger in its Hesperian garden, and enter its hospitable door.

When they asked, Whence came this transfiguration ? How is it that the wilderness of other generations now rejoices and blossoms like the rose ? the answer would be obvious, namely — From one seminal pistol-idea. Not from that idea as a guess, inchoate and half-produced in dreaming fancies; not as rudely developed in actual arms, where every advantage had a drawback; not as it dawned upon the boyish Colt in the Indian Ocean, nor even as perfected by a dozen years of experiments and experience, when, in the progress of improvement, complexity had yielded to simplicity, and delicacy to strength; but as utilized by means of concentrating on its fabrication a hundred varieties of witty invention, and by means of making it known throughout all the world, through as many varieties of shrewd advertisement.

As thus evolved—one success, the last result of countless failures—the pistol-idea is one of those which move the world. It first tamed wild nature, making a tract which had been "neither sea nor good dry land," henceforth terra firma. Then it incarnated itself into a colossal Armory, filled with more than a thousand machines, regulated by more than a thousand cunning workmen. It accumulated such a capital, and evolved such processes, new, secret, or patented, that, when its own patent had expired, it remained no less lucrative than before to its inventor. When conflagration in one hour ruined its appliances, that had cost a million of money, it has caused another structure, and other contrivances yet more exquisitely adapted for its purposes, to rise, like London two centuries ago, from the blackened ruins. It was the first step in a progress which revolutionized modern warfare. It speedily constrained the British government, for the first time in all history, to permit the introduction of its creations free of duty. In Stuart's felicitous phrase, coined within ten years of its perfected evolution, "from the snow-capt Nevadas on the Pacific to the blood-red plains of the Crimea, the mountains of the Caucasus, and the jungles of Hindostan, it *reports* the triumph of American skill, and *blazes* the fame of an American name." The world cannot afford to let it die; no deluge can drown, no fire consume it.

Such is the pistol-idea; but its slow evolution into a tangible and indestructible fact, and the many wonderful processes by which it is constructed, fit for defensive or aggressive purposes, will be shown hereafter. Here in this meadow, and the adjacent upland, the one grand result is patent to all men;—these fields, which only ten years ago were subject to spring and autumn floods, and to the ice-floes of winter, are now traversed by solid highways, and filled with the tokens and instruments of honorable labor, and the comforts and delights of domestic life. The beginning of all this territorial change was the building of the Dyke, within whose protecting arms all this portion of the city lies. Its achievement—requiring the coöperation of public bodies, and many individuals of seeming adverse interests—is the best illustration of Col. Colt's character and success, his indomitable energy, his immense will, his commanding decision, which, if they did not court difficulties and opposition, at least welcomed them when they came, as opportunities of effort, and occasions for the display of conscious power.

Such is Armsmear to-day—substantially as it was when death surprised its great master-builder in the midst of his unfinished enterprises—a princely estate, yielding an income, from the willing coöperation of science, industry, and art, which no planter ever wrung from sinews bought or sold, or prince in past or present time ever realized from serf or tenant attached to the soil. And its beautiful Homestead, the seat of generous and elegant hospitality, and Christian benevolence, will grow more and more attractive as time matures its trees and shrubbery, and art and skill harmonize the near and distant features of the landscape. In fixing and embellishing his

permanent residence—the Corinthian capital of the solid
structure of his great enterprise—Col. Colt evinced the
possession of tastes which his impetuous temper and busy
career had not led his best friends to give him credit for.

He selected his site so as to appropriate to himself
the beauties of a rare and varied landscape, of which
these domains, with all their embellishments, are none
too fair an accessory. Below the table-land, on the brow
of which are the mansion and ornamental grounds, lies
the beautiful valley of the Connecticut. From this ele-
vation the eye, sweeping down the cultivated slope, and
past the teeming village at its foot, ranges over an
area of some thirty miles, embracing all the best features
of the rural scenery of New England—

> Where Nature in her unaffected dress,
> Plaited with valleys and embossed with hills,
> Enchased with silver streams and fringed with woods,
> Sits lovely.

From Windsor on the north to Glastenbury on the
south, the river may be seen at intervals—its current
ever slow and lingering, as if reluctant to exchange these
placid scenes for the unrest of the sea. Skirting the
banks, and stretching away on either hand, are luxuriant
meadows, fertilized by annual alluvia, their broad reach
broken here and there by hazel copses, or dotted thickly
with ashes, elms, and maples, standing apart in conscious
grace, and occasionally brought into bolder relief against
the gleam of some passing sail. Beyond, to the east,
may be seen the cultivated farms of the upland, their
irregular patches and party-colored crops set mosaic-like
in the frequent openings; while here and there a tapering spire and occasional glimpses of white mark the site

8

of scattered villages, whose shaded streets may be traced
by denser lines of green. As the range of vision is
widened, these details of the scene, becoming less and
less distinct, are finally lost in the dense foliage—a vast
and continuous forest, as it seems, reaching to the far-
off hills whose undulating outline bounds the prospect.

To this fair landscape each season of our changing
climate contributes its peculiar beauties, affording to the
scene an almost infinite variety. Like a vast painting,
we may see it develop with the progress of the year.
Gradually the snowy canvas, on which bare trees and
scattered houses stand out like etchings, taking on at
first a dull brown preparation, shades into a ground
coloring of green. One by one each rugged feature
grows into softer outline. In endless succession the
different varieties of vegetation take their appropriate
forms and places, and under the patient handiwork of
nature are matured into finished detail. In the bright
coloring and minuteness of the foreground—the less
definite outline and subdued tints of the middle distance
—the mellow shadings of the hills, ever varying as the
pale green of maples, the darker hues of oak and ash
and beech, or the sombre tints of evergreens in turn
prevail—are elements of the most agreeable contrasts.
At every change from morning to evening, from sunshine to cloud, from summer calm to sudden storm and
tempest, we are surprised by new effects of light and
shadow and perspective. And so, day by day, and
through the seasons, the picture grows in scope and
beauty and detail, till, finally, over the perfected composition is wrought "the many-colored wonder of the
autumn."

MAP OF
South Meadow,
HARTFORD, CONN.
Showing Lands purchased by
COL. COLT.

THE CONNECTICUT RIVER FLOODS.

THE extensive and destructive character of the "Fresh-ets" of Connecticut River, from which Col. Colt under-took to protect, by an encircling causeway, a princely domain of three hundred acres, to be covered with massive structures for workshops, warehouses, and dwellings, all to be erected at his own cost and risk, will be seen from the following notices from contemporary papers of two of the highest floods on record.

The *Weekly Courant* of March 23, 1801, thus notices the spring freshet of that year, which has been since known as the "Jefferson Flood :"—

"By reason of the heavy rains during the week past, the streams in this part of the country, as far as we have heard, in every direction, have arisen to an unex-ampled height, and caused an immense destruction of property, public and private. Bridges, mills, fences, build-ings of all descriptions, dwelling-houses, barns, &c., &c., are swept away; many families are reduced to distress, by either being driven from their habitations for a season, or in having them completely destroyed. In this town there is no mark of so high a flood since the year 1692. The rise was so rapid, that people were scarcely able to secure their most valuable property in stores and houses, before the buildings were filled with water. Every family in that part of the town which lies near to the river, has been forced to flee for refuge among their neighbors; many of the families were taken from the windows in boats and carried away; some of the one-story houses are in water to the roofs; the lower

stories of all are nearly filled; and the scene wears the
appearance of extreme desolation and melancholy. Front
Street, from the Little River to the North Meadow gate,
is so full of water that boats pass the whole length. At
New Hartford we are informed that one man was
drowned. In this town we have not experienced the loss
of any lives."

The *Courant* of the subsequent week contains a letter
from West Simsbury, which states that on the Farmington River and its branches there were destroyed seven
grist-mills, five saw-mills, fourteen bridges, two clothiers'
shops and works, one dwelling-house, two barns, and
immense quantities of timber and fencing.

The same paper, now become a "Daily," on Monday
morning, May 1, 1854, under the caption "Great
Freshet. The water is higher than in the Flood of
1801 !" has the following editorial:—

"This city is now in the midst of the greatest freshet
which has ever visited it. We say the greatest freshet,
for as we write (ten o'clock Sunday evening), we believe
the water to be above the most reliable mark of the
flood of 1801—that on the old Distillery, corner of Tal-
cott and Front Streets. At seven o'clock the water was
two inches above the mark of 1801, at the toll-house;
at that time it was twelve inches below the mark on
the old Distillery. It was then rising about four inches
an hour.

"It commenced raining on Thursday evening, and since
that time has rained almost constantly, and very power-
fully. The water in the Connecticut was at that time
high and rising, and of course has since risen with very
great rapidity. It rose about as rapidly last evening

as at any time during the day, notwithstanding the surface the water covers must be very much greater than in the morning.

"All that part of our city lying contiguous to the Connecticut and Mill Rivers is of course flooded, and the North and South Meadows present to the eye nothing but one immense sheet of water. Some of the smaller houses on the banks of the river are entirely submerged, and many others nearly so. The inhabitants were taken away in boats.

"The large establishments of Messrs. Woodruff & Beach, Fales & Gray, and Col. Colt, will suffer heavily, not only in actual damage, but by a suspension of business, which will be inevitable. All the other manufacturing establishments and stores in Commerce and Front Streets have water on their floors or in their cellars, and must be damaged somewhat.

"The water covers all that part of the city lying north of Pleasant Street, and east of Windsor and Front Streets to Mill River. South of Mill River the water comes up into Sheldon Street almost to the blacksmith's shop, and of course fills the streets running parallel with Main Street. The South Meadow is entirely covered with water. Mill River is of course swollen not only by back water from the Connecticut, but by the immense quantity which it has received from the surrounding country, and the smaller streams by which it is fed. An embankment at Imlay's Mill was washed away, and the water forced a passage south of the mill, carrying from their foundations a small dwelling and a barn. The inmates of the dwelling had barely time to get into a boat with such clothes as they had on them. Lower down, the water runs through Ford Street, West Pearl, Hicks, and some other small streets in that vicinity—covers a part of Mill, Elm, and Arch Streets, and we believe is entirely over the new 'East Bridge.'

"Col. Colt had a large number of men at work in the meadow, who were living in shanties which had been erected there. The water rose so rapidly on Friday night as to make it necessary to take them off in boats on Saturday morning. They are now in some of the lofts near the river, and Col. Colt has made provision for their comfort.

"We have no particulars from East Hartford, but a large part of the town must be covered with water.

"During yesterday, one or two dwellings came down the river, but, so far as we could learn, no bridges or parts of bridges. It is reported, however, that several have been swept away on the Farmington River.

"'The Gully Brook Bridge' on Albany road is gone.

"The steamboat from New York arrived about five o'clock yesterday afternoon, and landed her passengers on the 'Great Bridge.'

"The merchants doing business near the river were busy removing their goods to places of safety—generally to the second story of their stores.

"The railroad tracks running from this city are doubtless more or less injured, but to what extent cannot of course be now known. One of the tracks of the Hartford and New Haven road, near Long Meadow, is somewhat washed, but the other is still good. It is reported that the bridge at the railroad junction near New Haven is gone. The track of the Willimantic road in the meadow is submerged, and will doubtless be injured.

"To add to our calamities, the water has so interfered with the gas-works that they were unable to give their usual supply last evening, and as the occupants of many dwellings, hotels, and public buildings, have disposed of their fixtures for lighting with oil, the annoyance is very great.

"The Mayor of the city, Hon. HENRY C. DEMING, was untiring in his exertions yesterday to see that every thing was done to rescue and relieve those surrounded by water. He ordered the City Hall to be warmed and opened, for the reception of such families as were drowned out, and directed boats to cruise through the night, to rescue those in danger. If the water continues to rise during this day (Monday), those of our citizens who have boats to hire, will confer a favor by communicating the fact to the Mayor, as he finds it extremely difficult to obtain any boats for the service of those in jeopardy.

"We have little intelligence from other places, owing to the interruption of the telegraph. A dispatch received from Springfield, at six o'clock last evening, says that the river there was twenty feet above low-water mark, and rising as rapidly as at any time during the day. The flooring of the Turnpike Bridge was twelve or eighteen inches above the water, but would doubtless be submerged. Several houses on the west side of the river, as well as a number of buildings along the banks of the city, were flooded. A boat had been running several hours, for the conveyance of passengers through the lower end of South Main Street. The dispatch adds, that the railroad track was damaged in one or two places in the city, and probably more seriously both above and

below. Another dispatch, to a gentleman in this city,
from the same place, and dated the same hour, says that
the river was six inches higher than in the great flood
of 1801, and rising three inches an hour.

"The flood of 1801, to which reference has been made,
was twenty-six feet two inches above low-water mark,
and was at its height on the twenty-first of March. A
flood occurred in 1692, when the water rose to the same
height. In 1843, the water reached the same height into
eleven inches. In 1852, the water was twenty-three feet
above low-water mark."

The *Times* of Monday evening, May 1, after recording
in three columns the destruction caused by the flood,
adds :—

"The flood in the Connecticut came to a stand at half-
past two, P. M., to-day (May 1), having risen above low-
water mark to the unprecedented height of twenty-eight
feet ten and a half inches, being higher by one foot
and eight and a half inches than any other flood during
the two hundred years that the town has been settled.
The highest floods recorded are as follows :—

"Flood in 1692 . . . . 26 feet 2 inches.
 "   " 1801 . . . . 27 " 2   "
 "   " 1841 . . . . 25 " 6   "
 "   " 1843 . . . . 26 " 3   "
 "   " 1852 . . . . 23 " 0   "
 "   " 1854 . . . . 28 " 10¼ "

"His Honor the Mayor has directed the City Engineer
to make prominent and lasting marks at various points,
indicating the precise height of this flood."

In the midst of these disasters, "the croakers" made

themselves heard, in the streets and in the public press, on the folly of attempting to thwart the purposes of Providence by building dykes. A "Tax Payer," in the *Times* of April 29, attempted to prove satisfactorily "that the whole deposit in the Connecticut Valley is permeated by the river, and is as open as a sieve. He should like to know how a dyke is to exclude the river, when it rises from below as rapidly as it flows in from above! The attempt to protect the South Meadows by embankments from the rise of the river is very like shutting the hatches of a ship down, to keep it afloat, when the bottom is knocked in." To this correspondent the editor replies: "It may be true enough that the work proposed by Col. Colt is difficult of accomplishment, but he is a practical man, and is willing and able to take the risk; and in place of grumbling about his project, we ought to allow him to take the risk, and the benefits of the risk, if there be any in it." Not content with this reply of the editor, Col. Colt was out in the next issue of the same paper, when the flood was at its highest, in a card, from which the following is an extract:—

"It is not very generous in 'Tax Payer' to take an opportunity for attacking me when all my attention is required to protect my property from the destructive flood which is upon us; but I am not yet discouraged, even by the combined forces of the land and water. 'The floods have lifted up their heads' higher than ever before; but that only proves the greater necessity of protection against them. Every year a scene like that of to-day is enacted, but we do nothing to guard property and life; and when any one citizen is willing to incur

9

the risk of contending with the river, such people as the 'Tax Payer' do all they can to defeat him. Now, sir, I am not afraid to face the music, or create it; and if the city of Hartford will agree to relieve my property from increased taxation, I will bind myself to exclude the river from the South Meadows; and, more than all that, if the city of Hartford will pay for it, I will agree to dyke the Connecticut River from end to end of the city, so that nothing less than Noah's flood can reach the houses which are now inundated. The circulation of the *river* in this city is not such a blessing that we ought to incur these heavy losses every spring to enjoy it."

It is worthy of remark, that the Council was author-ized by the city voters, December 27, 1855, to construct a dyke along the northern bank of Little River, and the western bank of Connecticut River, so as to protect the portions of the city heretofore visited for two hundred years from the devastations and inconveniences of similar floods, following in this the example of Col. Colt and his suggestion in the above card.

The hint of the possibility of a still higher flood than that of 1854 was not lost upon Col. Colt, who forth-with added another foot to the top of his embankment, to provide against the melting of deeper snows, or the fall of more abundant rains at any future time.

Since 1854, the highest freshets have been as follows: in 1859, March 21, twenty-six feet five inches; in 1861, April 18, twenty-one feet seven inches; in 1862, April 21, twenty-eight feet eight inches; leaving everybody and every thing dry and safe within the Dyke.

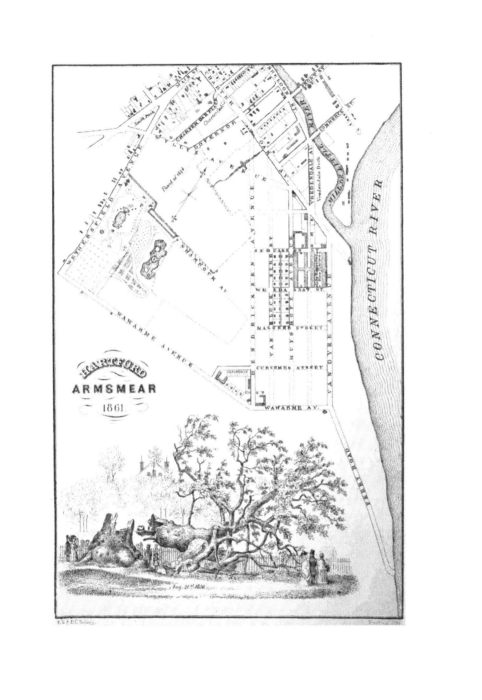

HARTFORD
ARMSMEAR
1861

CONNECTICUT RIVER

III. ARMSMEAR.

THE HOMESTEAD.

"Dost thou, Man, sigh for pleasure?
    Oh! do not widely roam!
But seek that hidden treasure
    At home, dear home!
There blend the ties that strengthen
    Our hearts in hours of grief,
The silver links that lengthen
    Joy's visits when most brief."

WHEN fortune, so long adverse to Col. Colt, began to smile, she lavished on him all her favors. Turkey, England, the whole length and breadth of his native land, all were before him, where to choose his local habitation—whether for labor or for rest. He had also seen them all. Yet he fixed his abode in the city of his birth.

In this he resembled Shakspeare, who invested his first earnings in a home at his native Stratford, and, when once independent, retired there to spend the Sabbath of his years. Nor is this course surprising. The magic of early associations grows stronger with the lapse of time.

"As when the sun prepared for rest
Has gained the precincts of the west,
Though his departing splendors fail
To illuminate the hollow vale,
A lingering light he fondly throws
On the dear mountain-tops where first he rose."

Besides, whoever has made a name in the world, and become a power, feels it most exultantly when near those once his superiors, but whom he has surpassed and over-topped. The misfortune is that there also is he most envied, and hence slandered and abused, just as Joseph was, while among his brethren, but not in the court of Pharaoh. There is also a *caritas soli*, a love of one's native soil not to be accounted for, but not to be denied. It is proverbial that the Swiss adventurer, having earned a fortune in the rich lands around, will return with an untravelled heart to spend it in the lap of his poor mother.

> "Such is the patriot's boast, where'er we roam,
> His first, best country ever is at home."

Had not this been human feeling, banishment could never have been judged among the most cruel punishments, nor would myriads have said with Cicero and Foscari, *potius vita quam patria carebo*, nor would poets have sung—

> " Dear the school-boy spot
> We ne'er forget, though there we are forgot."

We need not wonder, then, that Col. Colt, deaf to tempting proposals elsewhere, decided to establish himself in Hartford. Here he discovered such a site as he needed for his gigantic factories, and that where other men were as far from seeing it, as they were from ex-cogitating his pistol and his pyrotechnic ideas. How his dyke practically added nearly half a square mile to the territorial extent of the city, we have told elsewhere. Many lots adjacent to his Armory he bought up at prices then counted enormous, and offered more for others than they have at any time since commanded. No sooner had he thus become master of a suitable base of operations, than he began, though a bachelor, to build him

a house and to improve his grounds. When he became a family-man, and children were born to him, his
cares for domestic surroundings increased tenfold. Indeed,
he was never so active at adorning the paradise of his
home as when surprised by death.

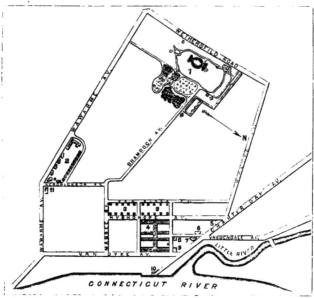

(1) Mansion and Grounds. (2) Potsdam. (3) Residences of Armorers. (4) Armory. (5) Charter Oak Hall.

The estate of Col. Colt was so laid out and has been
so labored as to take not a little advantage of its capabilities for picturesque effects. Its extent is about one-
third of a mile broad and two-thirds long. The shorter
face lies along Wethersfield Avenue, and the longer
stretches down to the Connecticut, including nearly the
same number of acres without as within the Dyke.

As we look forth from the mansion, upon its grounds,
the eye reposes upon the broad lawn, flanked on the left
by the deer-park, with nothing between but an invisible

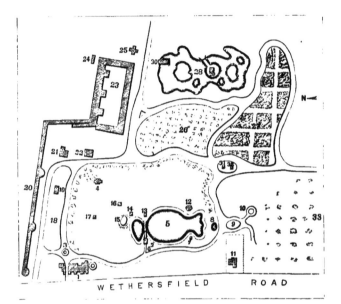

(1) Mansion. (2) Porte-cocher—Uffizi Dogs (3) Fountain—Boy and Swan. (18) Deer Park.
(28) Flower and Fruit Houses. (5) Lake. (15) Grove of Graves. (28) Swan and Duck Pond.
(26) Dwarf-Pear Orchard. (12) Elizabeth's Bower. (34) Charter Oak, Jr. (85) Kiss's Amazon.

fence, and, on the right, by a verdurous wall of flowers
that rise to shrubs and trees, closing the view street-
ward. Toward the south, this lawn is bounded by
clumps of trees, with glades between opening glimpses
of water;—toward the east, it stretches away unbroken
beyond the crest of the plateau or natural shelf as it
dips down to the river interval, leading the eye on,
above tilth, groves, and pastures, to the Armory and

Willow Works, half a mile distant, and so, still on, across the river to upland woods, and blue mountains where earth rises to heaven. The lawn, rolled and smooth shaven every fortnight, expands a velvet sward where the eye must love to linger. Near its edges it is studded with a bronze group, a statue, more than one vase, and more than a score of sentinel evergreens. Poured round all, in curves ever varying and rounding off all angularities, is the serpentine carriage-drive.

Approaching the grounds of Armsmear from the city, the eye meets first the little temple, as it were of Ceres and Pomona, which stands at the head of the very extensive green-houses. Across a broad expanse of shaded turf stands the homestead itself. A long irregular front lies parallel with Wethersfield Avenue; its lines well broken by the covered arch entrance into the grounds, the retreating façade of the dining-room, the oriel window, and the corner tower.
10

Violating, no doubt, many canons of architecture, and
wanting that unity of effect which the building would have
had, if it had been all erected at once, the house is nevertheless a very noble one, suggesting at once the elegance
and the refinement, the comfort and the privacy of such a
home.

His house, built, unbuilt, and remodelled at different
times, and with dissimilar aims, could not fail to be a characteristic type rather of the unique than of the unities in
architecture. Yet is it of massive stone, spacious, towered,
domed, with large halls for state occasions and crowds, as
well as cosy cabinets and boudoirs for household comfort
and genuine sociability, which, according to the classic
maxim, we taste oftenest in coteries neither less in number
than the Graces nor more than the Muses. Its cabinet of
memorials deserves, and shall have, a chapter to itself. The
numberless transparencies of porcelain biscuit, not hung
upon its windows, but set in the sashes themselves in place
of glass, are an embellishment of great rarity, and varying
its beauties with every change in the weather. In all its
appointments, this residence is one which everybody must
regret that the Prince of Wales did not visit, as he would
have done on his progress from New York to Boston, had
he not found it necessary to take the route through Albany.
From this dwelling the future sovereign of our mother
country would have turned with an impression we wish he
might have had concerning the material prosperity which
under democratic institutions was crowning a mechanic.

As the illustrations in wood from photographs bound
up in these pages will portray architecture better than a
volume of words, we shall spare our readers much of detail
about the residence. But the environs of the house are also

beyond description, and almost equally beyond engraving.
The coloring, the fragrance, the sounds, the varying charms,
the life, the motion, the vastness, how much can photo-
graphy render back to us of all these? Let those who can,
see with their own eyes. For such as cannot, it is well that
sun-painting and the pen do their little utmost to convey
an impression of landscape gardening on a liberal scale.

From the south, as seen across the lawn from almost any
point on the lake, the residence, with its massive tower on
the left, and the light oriental looking dome on the right,
and the airy pillars and pinnacles of the conservatory
between, with vases and statues, flowers and fountains,
children and animals on the lawn, has a very imposing as
well as domestic appearance. All views are best when we
look from the sun rather than toward it. Accordingly
this southern aspect is most brilliant on a morning walk.

The east view gives its best side as seen across the
smooth carpet of the sloping lawn, catching the airy pinnacles of the conservatory, the beautiful octagonal corners of
the little sitting-room, whose windows open on as fair a
landscape as the eye can sweep over anywhere, and the
long piazza, to whose steps the broad green lawn rolls up.

We enter between the Uffizi dogs, copied in Carrara
marble, which guard the gateway. Straining at their
chains, with open mouths they bay defiance against imper-
tinent intrusion, or deep-mouthed welcome to friends
returning home. We cannot but stop to admire their
vigorous proportions; the swollen veins and the starting
muscles; the splendid strength of chest and haunch that
almost live under the skilful chisel of the sculptor.

Through a narrow vista the eye takes in a picture
of surpassing beauty. A grassy knoll, crowned with a
gnarled and picturesque old apple-tree, beneath whose

shade is a little fountain of a boy and swan in bronze,
introduces us to the first of a graceful series of decora-
tions of this kind. This is a feature of the grounds at
Armsmear deserving notice. The *temper* of the architecture of the house is Italian; and though in the ample

grounds the stiffness of what Downing calls the "geometric style of gardening" is nowhere visible, there are yet indications of the combination of the English and Continental systems. "The verdant sculpture," as Pope in the Guardian calls the old French custom of trimming trees into fantastic shapes, gives place to the appropriate introduction, among the natural beauties of the landscape, of various classical accompaniments in the form of

vases, statues, and fountains, which, especially when as in this case the adornments are all in veritable bronze or marble, harmonize with and enrich this style.

> "Good rules of old discovered, not devised,
> Are nature still, but nature methodized."

Accordingly the attractions of these pleasure-grounds are enhanced by adherence to certain canons of landscape gardening. One is, that a vase should never, in the open air, be set down upon the ground or grass without being placed upon a firm pedestal, since without a base it has a temporary look, as if it had been left there by mere

accident.   Another is, that statues and all highly artificial ornaments should be stationed near the house.

In such a combination, each 'thing of beauty'—the
sculptured vase, the old stump, even in its decay made

beautiful by living flowers and trailing plants, the rustic
bridge, designed for ornament as well as use, and the
statues of Hebe, Apollo, and other classic characters in

marble, gleaming out of their green setting in sparkling
white, or figures full of life, like Kiss's Amazon, repro-
duced in bronze—adorns and completes the other, and
nature developed by cultivation is decorated by art.

Far different from these artistic adornments of the grounds are evergreens spoiled of their natural beauty by being cut into fantastic resemblances to men, animals, or events, of which Pope, in his essay on landscape gardening in the Guardian (No. 173), gives a list from a whimsical catalogue of a town gardener, who proposed to distinguish the villas and gardens in the neighborhood of the metropolis from "more barbarous grounds, where gross nature is followed," by the "imagery of greens." "He has arrived to such perfection, that he cuts family pieces of men, women, or children. Any ladies that please may have their own effigies in myrtle, or their own husbands in hornbeam. He is a Puritan wag, and never fails, when he shows his garden, to repeat that passage in the Psalms, 'Thy wife shall be as the fruitful vine, and thy children as olive branches round thy table.'" The following are specimens of his "green sculpture."

"Adam and Eve in yew; Adam a little shattered by the fall of the tree of knowledge; Eve and the serpent flourishing."

"The tower of Babel not yet finished."

"A quickset hog, shot up into a porcupine, by its being forgot a week in rainy weather."

"Noah's Ark in holly, standing on the Mount; the ribs a little damaged for want of water."

Sweeping round the broad gravelled carriage-way, our eye rests on the deer-park, shut in by an almost invisible fence, and offering to those graceful animals a safe and sheltered home.

> "Theirs are eyes serenely bright,
> And on they move with pace how light!
> Nor spare to stoop their heads and taste
> The dewy turf with flowers bestrown."

11

Tame enough to know no fear, yet with enough of
nature in them to be startled into the graceful beauty
of their poised heads or nimble legs by an approaching
footfall, they add living grace to the scene. We have
them in all their shapes. The branching horns of the
antlered buck, the soft, deep, tender eye of the doe, and
the pretty mottled coat of the young fawns, give us the
very incarnation of animal refinement and delicacy.

Our walk round the lawn keeps in full view the
green-houses, lining the whole length of the northern
side of the place, and then branching off into a large
quadrangle; outside and in, they are beautiful to look
upon. Cheating the seasons, and transplanting the tropics
to our colder clime, they repeat the Homeric picture of
the garden of Alcinous:

"The balmy spirit of the gale
Eternal breathes on fruits untaught to fail;
Each dropping peach a following peach supplies.

On apples, apples, figs on figs arise.
The same mild season gives the blooms to blow,
The buds to harden and the fruits to grow."

You walk through them in a bewilderment of perfume, that amounts to the lusciousness of taste. Here
the innumerable colors and combined fragrance of flowers
from every country ; there the narrow aisle leads between
a good-sized plantation of cotton, its pods bursting with
their woolly fruit, on the one side; and on the other as
large a crop of rice, whose bearded spears are swollen
with their pearly kernels. An opening door admits you

into the graperies, whose massive clusters of amethyst
and opal hang out from the fretting leaves and tendrils
that festoon the glass overhead; and so you pass through
house after house in which the homelier vegetables of the
kitchen-garden make summer seem to come in February,
until the pine-apples and bananas, and the rich-leaved

tropical plants in thrifty bloom and beauty, assert in a
new sense man's dominion over the green things of the
earth. Nor is the enjoyment of these things for the eye
alone, as many a gathering of guests, and many a home
of friends, and many a chamber of the sick and poor
bear grateful witness.

Col. Colt's love of flowers waxed stronger in him to
the end of his days. His flower-houses, begun as soon as
his dwelling, grew till they were popularly reported to
be a mile long, and in fact, if stretched out in one line,
they would extend no less than twenty-six hundred and
thirty-four feet. Of these about eight hundred feet are
green-houses, or, as some gardeners call them, cold-houses.
The rest are hot-houses. The greater portion of them
forms a hollow square. There are some span-roofs, though
the lean-to variety is more common. There are four

octagon rooms rising in steeples, comfortable with willow
chairs, and garnished with a wheel-window, colored red,
yellow, and blue.

Thirteen of the apartments are graperies. Among the choicest vines are the black Hamburg, the Barbarossa, the Sweet-water, the Muscadine, and the Chasselas. The annual produce is at least a ton, gross weight. Some of the clusters remind of the *Boumastus*, or cow's udder—the largest variety known to the ancient Greeks. But we still look in vain for any specimens of the millennial vintage, when, the rabbins say, each grape will be so grandiose that there will be no need of wine-presses or of treading, but only to provide a faucet and thrust it into a single grape, for getting all the pure juice one can drink. Still, it may be doubted whether any rabbin will promise us the fruits of the vine ripening throughout more months of the year than they here ripen. Some are forced on till they are sweet in April; others are kept back, and so snubbed that in November Æsop's fox would pronounce them still sour. Strawberries in like manner come precociously on the table in February, and cease not to grace it till those growing wild are gone. Four thousand pots bear about four hundred quarts. Cucumbers are ready for use at merry Christmas. Nor do they fail to be all the winter long.

The only drawback on such unseasonable dainties is that they confound our ideas of chronology. Most people date from the coming of first-fruits, rather than from any fast-days, or saints' days. Having all fruits always, they may forget that seasons ever change.

The fig-house, or *Ficetum*, boasts a better article than inspired Byron's couplet:

> "From a Tartar's skull they had stripped the flesh,
> As you peel its fig when the fruit is fresh;"

for the figs grown here have scarce any peel, but are

almost all a pulp which, melting like kisses, would extort
praise even from that prince of prophetic grumblers,
Jeremiah. The crop is near two thousand fruits.

Peaches and nectarines, though dwarfed, are ready to
burst through the roofs. They pay a tribute of seven
thousand fruits.

In the Pinery, we survey half a hundred pine-apples
still nurslings, as many half-grown, and an equal number
just maturing. Strangers to the habits of these illustrious
foreigners will be grieved to hear that each plant bears
no more than one apple, and that but once. Yet it may
boast with the lioness, "If I bear but one, it is a lion."
What a contrast to this single product we see in the
prolific banana. One of these now in the green-house
has a stalk two and a half feet around, and holds two
hundred and thirty fruits, piled up, as it were, on its
single spike.

Flowers please one less of our senses than fruits, and

are hence harder to speak of. Yet whoever has seen one rose, may believe that the Lamark and Banksia are here transcendent. She who has watched the unfolding of one camellia, would never cease to admire the forty varieties here. But the orchids, hung in the air yet grovelling earthward as they grow, thrusting out pale flowers through the crevices of their log cabins, are too fantastic to be credible to him who has not seen them all. They ought to be called idiots or insane, so crazy are they in their pranks. One of them, however, is a sacred freak, the *Peristeria elata*, or Holy Ghost flower, shaped like a white dove with wings outspread. The cactuses are no less misshapen and unfinished than we should expect to see them, had they been made, as some say the world was, by a fortuitous concourse of chaotic atoms.

There are no travels at home so impressive as a walk among flourishing exotics. What most strikes a New Englander in Southern Italy, and convinces him that he is far beyond the sea, is that the robes of nature, the whole aspect of the vegetation, are so unlike what he sees at home. Similar are the impressions in this greenhouse, and they intensify the farther one walks. Nothing in the tropics can be more high-colored than the *Dracæna nobilis*, transparent as painted glass and flame-bright all over. Time would fail us to speak of hydrangeas, the stephanotis, and datura brakmanza; of Eastern acacias, fuchsias, caladiums, hoyas, calceolarias, and, above all, azaleas and late-flowering chrysanthemums, multitudinous and of manifold variegations. Such growths of more genial climes brought back to Col. Colt the scenery of the gorgeous East, no less vividly than if he had brought home one of its sunny isles in his sailor-boy pocket. Some

loungers in these narrow rooms displaying nature's whole
wealth, as they gaze on sections cut out of the ends
of the earth, brought safe hither and expanding as if in
a purer ether, a diviner air, love to think of the legend
that when the palm of the Egyptian desert bowed down
its head for offering dates to the child Jesus, he broke
off a leaf of it and sent it to heaven by one of the
angels who ministered to him, and that, there planted,
it is becoming the palm-grove of the skies, *cœleste palmetum*,
from which the ransomed of the Lord will pluck the
branches they shall wave in homage as they stand before
His celestial throne. .

The hot-houses, so far as they do not spread the table,
culminate in the CONSERVATORY, a show-case worthy to set
off the selectest gems which Flora can present. It is one-
third as long as the Astor House, and its width is
two-thirds of its length. It is composed of glass panels—
each usually two yards long, set in frames of iron fashioned
into foliate arches. Some of them are red, yellow, and
violet. At each corner is a sort of minaret, while the
centre rises in a dome capped with a golden pine-apple,
as well as flanked on each side by two domelets.

Beneath the central dome and in the midst of the
central front, a bronze Triton, upward man and downward fish, blows up a triple water-jet on high, which
sparkles into rainbows in the sunbeams, and drops in
diamonds into an ample laver, freshening it for the gold-
fish who gambol on the gravel or lurk beneath the sea-
god. The floor is of marble tesselated in red and white.
Six iron pillars sustain not only the roof, but ivy and
other climbers.

The whole centre is filled by an iron pyramid, on the

steps of which the flower-pots are ranged,—the *élite* of
the green-houses,—brought day by day to this post of
honor, each when its flower is at its acme, to be the
perfume and suppliance of a minute. It is nothing less
than one gigantic bouquet, culled, as it were, from all the

continents,—from all their fields and all their gardens.
Humming-birds, those flying flowers, resort from the four
winds to this " box where sweets compacted lie," flitting
through the transom windows, or the four wickets which
open upon the garden. The bird seen in the engraving
is one which in fact lighted on the pavement just as our
artist was taking his sketch. The floral pyramid, renewed
all the while from the hot-houses, as by a stream from
the cornucopia, is as unfading as the Roman table in the
Borghese palace, which is inlaid with a specimen of every

12

known gem. Both are ever fair and ever young. But
the table becomes monotonous, being always the same,
while the variations in the conservatory are exhaustless,
and new every morning.

> "Taste after taste upheld by kindliest change;
> Yet all so forming a harmonious whole,
> That, as they still succeed, they ravish still."

It adds to the witchery of this spot that it has an
excellent outlook far and near; that glass doors open into
it from the garden, from the porches front and rear, from
the library and parlor back of it; that its light and heat
may be tempered at will by blinds, and that, of an
evening, it may be illuminated by half a dozen chandeliers, all the flames shooting out of porcelain flowers.
On the whole, viewed from within or from without, as a
whole or in detail, by day or by night, the conservatory
—with its cooling, soothing plash of waters, its rich dis-
tilled perfumes, its golden fruits and snow-white blossoms
on the same bough, its leaves of velvet, and its flowers
in color, figure, shading, grouping, ideals of beauty—seems
in our work-day world the dream-land of poetry, and
realizes "the stately pleasure dome" which Kublah Khan
decreed in Xanadu.

But we tire of in-door life even in conservatories. We
feel it an injury and sullenness against Nature not to go
forth and participate in her rejoicing with heaven and
earth. Hence, in the progress of civilization, originated
the art of pleasure-gardening, and hence Col. Colt was no
less solicitous about his grounds than about his house
and his green-houses.

The gardens of Semiramis were classed among the

seven wonders of the ancient world. Solomon tells us that he made himself gardens. Indeed, primeval Eden was a garden. According to Milton, it was

> "A happy rural seat of various view,
> Flowers worthy of paradise, which not nice art,
> In beds and curious knots, but nature boon,
> Poured forth profuse on hill and dale and plain."

But, whatever were gardens in poetry, in practice they were either adapted only to culinary purposes, or ridiculously artificial, so recently as the opening of the last century. Macaulay's account of the abode of our first parents, as engraved in old Bibles, is a fair representation of them all. "We have an exact square, inclosed by the rivers Pison, Gihon, Hiddekel, and Euphrates, each with a convenient bridge in the centre, rectangular beds of flowers, a long canal, neatly bricked and railed in, the tree of knowledge, clipped like one of the limes behind the Tuileries, standing in the centre of the grand alley, the walks twined round it," &c., &c.

Artificers of such mathematical pleasure-grounds provoked some one to the ironical epigram :—

> "Soon to thee shall Nature yield
> Her idle boast.
> Her vulgar fingers formed a tree.
> But thou hast pruned it to a *post*."

Extremes, however, meet, and so the reaction from this artificiality swung the pendulum to a naturalness equally overstrained—nature too severely true. Thus it came to pass that, in the generation when Rousseau was declaiming, and that far more effectively than Cooper's novels have done in our days, about savages as the highest ideal of human development, a garden, laid out in

accordance with the natural method, became confusion
worse confounded—

> " Where all the husbandry doth lie in heaps,
> Corrupting in its own fertility.
> The vine, the merry cheerer of the heart,
> Unpruned dies; the hedges, even-pleached,
> Like prisoners wildly overgrown with hair,
> Put forth disordered twigs; the fallow leas,
> The darnel, hemlock, and rank fumitory
> Do root upon, while that the coulter rusts
> That should deracinate such savagery.
> The even mead, all uncorrected, rank
> Conceives by idleness, and nothing teems
> But hateful docks, rough thistles, kecksies, burs;
> So that the vineyards, fallows, meads, and hedges,
> Defective in their natures, grow to wildness."

To call such studied neglect, or mimic wildness, gardening, was a misnomer no less egregious than that of
the artist who painted a dead wall or barn door of Spanish brown, and called it "The Israelites crossing the Red
Sea." When the painter was asked, "Where are the
Hebrews?" he said, "They have all passed over;" and
when the question was, "Where are the hosts of Pharaoh?" his answer was, "They are all drowned." All
the people who would have given human interest to the
picture had gone—either over or under.

In process of time it became clear that a garden, no
less than a picture, must have human interest, and landscape gardening was born, as it were a daughter of landscape painting.

A double error sometimes sets us right. It proved
so in gardening. The counter claims of the ultra-natural
and the ultra-artificial led to a compromise where both
blend in harmony, and it remains forever doubtful which
owes the more to the other.

Modern gardening, then, is a "modest art that secretly seeks to hide itself,"—or art's pretty comment on nature's pleasant text,—or, best of all—

"There is an art which with great Nature shares;
Yet Nature is made better by no mean,
But Nature makes that mean; so o'er the Art
Which, we say, adds to Nature, is an Art
That Nature makes;—that art itself is nature."

A landscape gardener superior to all before his time, and, if we believe Hugh Miller, to all men since, was Shenstone, a little more than a century ago. "It was his," says Johnson, "to point his prospects, to diversify his surface, to entangle his walks, and to wind his waters; which he did with such judgment and such fancy, as made his little domain the envy of the great and the admiration of the skilful, a place to be visited by travellers and copied by designers." In these Leasowes of to-day we see strong confirmations of their gardener's remark:—"What we build begins at once to decay; what we plant begins at once to improve."

There is perhaps no feature more striking and taking in landscape gardening, than what is called the management of water. Readily made offensive by stiffness in form or by stagnation, when rightly arranged it adds vastly to the comfort, the pleasure, and the beauty of an estate. A well-appointed lake, whose springs and conduits keep it clear and full, well stocked with fish, offering its mirror to the summer moon, and adding coolness and freshness to the summer wind, inviting one to profit by the exercise, or enjoy the easy motion, of boating in the warm weather, and skating in the cold, is a point of special, constant, and ever-varying attractiveness.

The focus of greatest interest in all the grounds is
about the lake. The notes of busy life, in the street
west, come softened there. Southward the outlook is
upon ornamental trees; beyond them a broad orchard;
and then, stretching away beyond the dyke, the open
spaces of the meadows, with noble trees, single or in
groups; and finally, out of the mass of foliage, the sharp
steeple of Wethersfield church.

The point of most perfect vision in this direction is
near the lake-side and vine-clad summer-house, which
its builder loved to call Elizabeth's Bower. The witch-
ing time, when all the loveliness of the landscape, so soon
to fade from our sight, is seasoned by special seasonableness, is evening, when the shadows fall long adown

the eastern valley. Then and there are the time and
place to thrill with the emotions which made Byron
write:—

> " When the last sunshine of expiring day
> In Summer's twilight melts itself away,
> Who hath not felt the softness of the hour
> Sink on his soul, as dew along the flower?"

Northward the prospect is of city spires bosomed high
in trees, while eastward it is diversified and far-reaching

even to mountains, which provoke imagination to expand
the vast to the infinite. From this point, too, the aspect
of the residence is decidedly oriental, owing not merely
to the gold-tipped and dazzling dome, and its pendent
pinnacles of the conservatory, but to the tent-like awnings,
the pointed arches, and the dome-capped tower in which
the house rises at its garden corner. Here, also, thanks
to flowers of all hues, as well as sombre evergreens and
gray willows edging the bright lawn and silver waters,
"the magic of color" is no unmeaning phrase. Moreover,
the lake-side stroller, or he who sits in its summer-houses,
or on the seats beneath the shade, finds a feast for his
eye in the Willow Works of broad-brimmed roofs, the
quadruple lines of the Armory—the real fountain from
which the whole garden is a stream—the peaked glass
roofs of the green-houses, and the islets in the pond
beneath the hill, and in the distance the ever-varying
aspect of the wooded uplands.

Nor is the lake itself devoid of attractions. Although
in part artificial, it was clearly intended by Nature for
this spot, as she showed by filling it with springs. The
central water sheet is three hundred and eighty-eight
paces in circumference. The banks are filled well-nigh to
the brim, and the waters are in places thirty feet in
depth. In part supplied by local springs, they are so
clear that we count many a grain of gravel far down in
their depths, and scrutinize the minutest movements of
the gold-fish, or Chinese carp, a score of which are often
in sight at once. Black bass are also abundant and large.
A row-boat is often launched on the lakelet, which affords
ample sea-room for an inexperienced oarsman. Swans,
however, plough its surface most frequently, and seem to

exult in their superior swimming, "smooth-sliding without
step." In a thicket where the ground falls off beyond
the triple water the swans had their nest, and, when
there were cygnets in it, rushed forth against passers-by
as fiercely as Hector when protecting his household gods
could have braved the Greeks. Sometimes, the waters
rush up in jets from two water-nymphs and a bronze
colt standing in the midst of the waves; but, when quiet,
they form a matchless mirror doubling the swans, the
statuary, the trees, and all the surroundings. The whole
outline, though irregular, has some likeness to that of
an Etruscan vase. A rustic bridge unites two projecting

points. Evergreens, which Humboldt styled the oases of
cold climates, and many other trees, shade the shores.
There is one of the weeping-willows which, drooping its
boughs into the water, might have given rise to a tradition
like that concerning one beside the Euphrates, which, when
Alexander in his boat was passing beneath it, caught the
crown from off his head—an incident that was interpreted
as an omen of his speedy death.

The very bold terracing into which the lawn rises on
the south leads us naturally thither, to find the most artistic and exquisite effects of this whole landscape. The

grouping here is perfect. Looking from the lower edge of
the lawn, the silver shimmer of the larger lake breaks on
the eye; the graceful rustic bridges, the fountains, the
summer-house—not put there for effect, but there because
some resting-place was needed from which to enjoy the
view—and the bits of statuary standing out against the
rich green; all these, fitting in each with the other, make
a *coup d'œil* of rare beauty; while a little off from the
"willows by the water-courses," the darker evergreen of
the cypresses and firs marks the sleeping-place of that
sacred dust, which is garnered here in the choicest spot of
the noble estate.

The steep declivity of the terrace which ends the lawn
shuts out of view from the house what we can see here—
the sort of every-day home of the water-fowls, whose more
dignified representatives are honored also with a more

13

elegant residence nearer the house. Here the two swans admit companionship of the inferior water-fowls, and allow the presence of their own offspring. Curiously enough, they claim the monopoly of the upper lake. During the hatching season nothing can exceed the faithful patience of

the mother, nor the father's courageous devotion. And during the gray gosling period, the "ugly ducklings" paddle in awkward circles round their elegant progenitors. But after that time, they claim sole and exclusive possession of the upper lake, where they "float double, swan and shadow," the livelong day, arching their pliant necks and ruffling their snowy wings, the very pictures of proud, contented, graceful elegance.

> "The feathered tenants of the flood,
> With grace of motion that might scarcely seem
> Inferior to angelical, prolong
> Their curious pastime.
> They tempt the sun to sport amid their plumes;

They tempt the water. or the gleaming ice,
To show them a fair image; 'tis themselves,
Their own fair image. on the glittering plain
Painted more soft and fair."

Indeed, they claim sometimes a monopoly beyond their
share. And the narrow walk around the lake is the scene
sometimes of rather ludicrous encounters, the intruder often
beating a reluctant and ignominious, but anxious and rapid
retreat from the ungainly figure that waddles after him,
with a power of mischief not to be mistaken, and a purpose
of mischief in his wing which not many are anxious to
test.

Whether we stand upon the shore, or view the lakelet
from the house, or steal side-glances at it as we ramble,
or look down upon the island pond, we still feel that a
landscape devoid of water must remain as imperfect as a
face without eyes. "Water," in the words of Kemp, " as
a source of variety cannot be overrated. The sparkling crystallizations or feathery spray of a fountain or
cascade; the ripple of a pool, as it is agitated by winds
or disturbed by fish; the reflections of lawn, plant, and
sky, which are so softly mirrored on its glassy surface
after a warm rain; the murmur, and music, and life of
a stream; the transparency, the glitter, the coolness almost
inseparable from the possession of water in any form;
are all causes of a well-nigh endless variety. And if
water-fowl can be encouraged, its liveliness will be far
more conspicuous."

Nor can even the casual visitor fail of figuring to himself, had Col. Colt lived to complete the Artesian well
which was interrupted by his death, and had it, like one
just sunk in Chicago, yielded half a million gallons a
day, what stupendous sheaves of foam would here be

now leaping up in everlasting life, music to the ear, dropping diamonds to the eye, and freshening the sultriest summer noontide for all the senses.

Few who saunter on the lake-side can fail to be interested in its marble statuary : the goat suckling her eager

kid ; the magnanimous mastiff teased by a puppy ; Hebe pouring out nectar for the gods, an office, some say, she lost for telling their convivial secrets, while others hold that she gained a better by marrying one of them ; but,

above all, the Apollo, the great glory of the Vatican, by
many pronounced the most peerless legacy of Grecian
sculpture, and so rapturously described by Byron:—

"Or view the Lord of the unerring bow,
  The God of life, and poesy, and light—
The sun in human limbs arrayed, and brow
  All radiant from his triumph in the fight.
The shaft has just been shot—the arrow bright
  With an immortal's vengeance; in his eye
And nostril beautiful disdain, and might,
  And majesty, flash their full lightnings by,
  Developing in that one glance the Deity.

"But in his delicate form—a dream of Love,
  Shaped by some solitary nymph, whose breast
Longed for a deathless lover from above,
  And maddened in that vision—are expressed
All that ideal beauty ever blessed
  The mind with, in its most unearthly mood,
When each conception was a heavenly guest,
  A ray of immortality—and stood
  Star-like around, until they gathered to a god."

Not far from the lake, though on the edge of the lawn, and, as seen from the house, contrasting with a background of dark foliage, stands Diana, with the fawn as her attribute. Her attitude is as when surprised while bathing in the vale of Gargaphia by the unlucky hunter Actæon. Ovid thus tells the story:—

> "Now all undressed the shining goddess stood,
> When young Actæon, wildered in the wood,
> To the cool grot by his hard fate betrayed,
> The fountains filled with naked nymphs surveyed.
> The frighted virgins shriek at the surprise,
> The forest echoed with their piercing cries:
> Then, in a huddle round their goddess pressed,
> She, proudly eminent above the rest,
> With blushes glowed; such blushes as adorn
> The ruddy welkin or the purple morn,
> And though the crowding nymphs her body hide,
> Half backward shrunk, and viewed him from aside.
> Surprised, at first she would have snatched her bow,
> But sees the circling waters round her flow;
> These in the hollow of her hand she took
> And dashed them in his face, while thus she spoke:
> 'Tell, if thou canst, the wondrous sight disclosed,
> A goddess naked to thy view exposed.'
> This said, the man began to disappear
> By slow degrees, and ended in a deer."

In the wilder and more distant parts, several stumps are mantled and decorated by flowers, which seem striving to hide the devastations of age and woodmen.

Nearer the house, young Charter Oak is putting forth a luxuriance of leafy honors to which our engraving by no means does justice; a century-plant, that looks already old enough for flowering, stands before the central swell of the conservatory, and foliage-flowers flame in the white vases. Art, too, is here most conspicuous, for the finest of the bronzes is seen in pride of place on the brow of a grassy mound, and above a pedestal of Port-

land stone. A fair lady on horseback is always the incarnation of strength and beauty rejoicing together.

The group here is triple, namely, the equestrian Amazon, who is the classical Joan of Arc, armed with a

dart, in mortal combat with a panther that has caught
her steed's neck in his massive jaws. The battle still
hangs in even scale, yet the burin of the artist would
fain embody mind victorious over brute force. The
greatest energy of motion is here expressed in the wri-
thing horse, the raging wild beast, and the vengeful virgin.
The original, by A. Kiss, which is twelve feet high,
ornaments the stairs in the Berlin Museum. It was
that artist's first masterpiece, and gained a first prize
in 1851 at the London Art-Palace. Its model was so
much admired in Germany that contributions for exe- ·
cuting it in bronze were taken up even in the churches.

The carriage-drive begins at the street-gate. Starting there, we pass under an arch, between marble dogs
which keep watch on either hand, copies of those that
do the same duty in the inner vestibule of the *Uffizi*
Palace at Florence. Before us, as we look through the
coach-arch, swells a grassy hillock beneath a patriarchal
apple-tree, and crowned by a bronze boy with his arm
round the neck of a swan—copied from an original by
A. Kalide. A critic of some eminence says: "It is a
charming group, of natural gracefulness. Nothing is
more simple and appropriate to be used as a fountain
ornament." Such is its use here. To a spectator from
the street-gate, the coach-arch serves as a picture-frame to
set off the hillock, the bronze, the fountain, and the tree.
On riding in, we may bear to the left around the deer-
park, where half a dozen bucks and does, with their
fawns, are skipping about as if unconscious of any
restraint upon their liberty. Their wire fence looks more
like a veil than a barrier. Once one of them escaped,
but after some months of wandering was caught, appa-

rently on his way back. Peacocks perking their sapphire
necks, and spreading their trailing trains, are at large on
the lawn or fly over into the park. We only wish their
voices were in keeping with their plumage, as legends
say they were in paradise. Sweeping past stables at first
hid by a screen of trees, green-houses, gardeners' quarters, and an orchard at the foot of the slope, we pass a
water-sheet with four or five islets, formed by damming

a brook. After winding round the extensive kitchen-garden, the drive forks, one branch leading to the Dyke,
while the other, passing a pair of marble lions, meanders
up to a grove of ornamental and fruit trees upon the
upper terrace, and thence, alongside the triple lakelet, till
it reaches the lawn again. From the verge of the terrace
there is a prospect over the flat below, corn-fields, meadows, grazing ground, not without monumental elms,
14

arching their plumes in many a feathery spray; beneath them, cattle feeding or reposing.

In all our rambles about these pleasure-gardens, we see so much done, and that so well, and in so brief a space,—"turning the accomplishment of many years into an hour-glass,"—that our imaginations are fired. We thrill with conceiving what greater miracles would have here been wrought, had their proprietor filled out his three-score years and ten, all the while increasing equally in resources, and in well-considered labors to develop to the utmost the picturesque capabilities of his possessions. Then, in the perfecting of this paradise, there would have been such a harmonious interaction of man and nature as could be described only by these words, which no man can mend, regarding a well-matched marriage:—

> "He was the half part of a blessed man,
> Left to be finished by such a she;
> And she a fair divided excellence,
> Whose fulness of perfection lay in him,
> But two such silver currents, when they join,
> Do glorify the banks that bound them in."

The result would have been a wondrous realizing of Tasso's romantic dream concerning the fairy-land of Armida, which, it is worth noting, was also chronologically the first ideal ever painted of a true garden. His words are:—"Every thing that could be desired in gardens was presented to the eye in one landscape, and yet without contradiction or confusion,—flowers, fruit, water, sunny hills, descending woods, retreats into corners and grottos; and what put the last loveliness upon the scene was, that the art which did all was nowhere discernible. You might have supposed (so exquisitely was the wild and

cultivated united) that all had somehow happened, not
been contrived. It seemed to be the art of Nature herself; as though in a fit of playfulness she had imitated
her imitator. But the temperature of the place, if nothing
else, was plainly the work of magic, for blossoms and fruit
abounded at the same time. The ripe and the budding
fig grew on the same bough; green apples were clustered
upon those with red cheeks; the vines in one place had
small leaves and hard little grapes, and in the next they
laid forth their richest tapestry in the sun, heavy with
bunches full of nectar. At one time you listened to the
warbling of birds, and a minute after, as if they had
stopped on purpose, nothing was heard but the whisper-
ing of winds and the fall of waters. It seemed as if
every thing in the place contributed to the harmony and
sweetness. The notes of the turtle-dove were deeper than
anywhere else; the hard oak and the chaste laurel, and
the whole exuberant family of trees; the earth, the water,
every element of creation, seemed to have been compounded but for one object, and to breathe forth the
fulness of its bliss."

From this sweet dream of the poet, we turn to take
one more glance over this fair landscape, this veritable
creation of a gifted mechanic, to which each season lends
a peculiar charm. The silver thread of the Connecticut
winds in and out under the encircling belt of hills whose
uneven horizon line melts into the sky. Here and there
among the many shaded greens of these well-wooded
hills peep out the clusters of villages, the spires of
churches, or the bits of cultivated land. Within this
boundary every thing tells of the founder of the estate.
The waving osiers of the Dyke, the sturdy buildings of

the Armory, the comfortable homes of so many of the
workmen, the pleasing fancy of the Swiss village, with
the carved work and uncovered timbers, and the external
staircase of the houses, and the broad, quiet meadow, so
rich and luxuriant in grass and noble trees, all these are
the enduring monuments of the greatness of Col. Colt's
character and achievements. And still the eye contracts
its gaze to dwell with ever new delight upon the more
artistic cultivation of the estate itself, the velvet carpet
of the lawn, the admirable grouping of the trees, the
lake's glittering repose, the sweep of the broad avenues,
the sculpture, the fountains, the flowers, until it falls
with loving reverence upon what seems the heart of this
great system, where that which was mortal of the man,
to whom his city and his country owe so much, is laid
away to sleep. How strangely do the mental creations
of the mind outlive the mind that gave them birth—such
perishable, fragile things surviving what seems so strong,
so living, so beyond the reach of death. But the strange-
ness is only in the seeming. The physical things, that
stay behind upon a longer lease of life on earth, end
where they began; "of the earth, earthy," and so to
perish with the earth. They are left here only as the
cocoon still lies upon the ground, when the brilliant butterfly wings its rainbow way towards heaven, fair symbol
of the resurrection hope. And that which is passed away
from earth is, after all, the changeless, the complete, the
immortal.

THE GROVE OF GRAVES.

" On, on, in one unwearied round
  Old Time pursues his way;
Groves bud and blossom, and the ground
  Expects in peace her yellow prey :
The oak's broad leaf, the roses' bloom,
  Together fall, together lie,
And undistinguished in the tomb,
  Howe'er they lived, are all that die.
Gold, beauty, knightly sword, and royal crown,
  To the same sleep go shorn and withered down."

Close by the lake, but embosomed in so thick a clump
of trees that they might not be noticed by the stranger,
are the family graves. Here lies the father, with his son
and two daughters. The spots of burial are marked by
simple slabs, and one marble niche which canopies a boy
among flowers gazing at a butterfly, the classical emblem

of resurrection. The sleepers lie with floral crosses and
hearts daily renewed upon their graves, beneath willows
that weep, and spruces of a verdure that never dies, all
circled by a hedge of the "tree of life."

The beautiful figure in the marble niche was modelled
from life by Bartholomew, on his last visit to his native
State, for a bust of Col. Colt's eldest child, to be placed
in his private room, among the choicest memorials of
affection. But the child dying soon after, it was executed in marble by the artist himself, who did not, however, live to complete the monument of which it consti-
tutes so appropriate a part. This child's first birthday,
which opened under other skies than ours, was thus
commemorated by Mrs. Sigourney:—

SAMUEL JARVIS COLT.

Born February 24, 1857.—Died December 24, 1857.

Lo—thy first birthday,
Baby most fair,
Cometh in wintry grace,
Cometh with smiling face—
Where art thou? Where?

Not on the mother's breast,
Loving and mild;
Not 'neath the father's eye,
Watching him tenderly,
Sports the brave child;—

Not in his nurse's arms
Roams mid the flowers—
Not in his cradle rich,
Carved out with jewelled niche,
Sleeps through sweet hours;—

Not in this mortal clime
Where sickness pines;
Look—o'er the realms of tears
Where, with no date of years,
Full glory shines.

There, scattered oft the earliest of the year,
By hands unseen, are showers of Violets found,
The little redbreast builds and warbles there,
And fairy footsteps lightly print the ground.

So—not like sorrow-bowed
  Pilgrims of time,
Here—in probation dim,
Dare we to speak of him,
  Being sublime!

Fond Love—that mourneth,
  Rest thee in peace!
Honor mid all thy pain!
Yonder bright angel-train—
Choir of the heavenly plain
  Thus to increase.

Side by side sleeps his little sister, whose departure was thus tenderly remembered by Mrs. Sigourney:—

ELIZABETH JARVIS COLT.

Born February 22, 1860.—Died October 17, 1860.

A bird!—a bird!—of sweetest song,
  Bright eye, and plumage rare,
A blessed visitant to make
  This world of ours more fair.

Its nest was 'neath the palace eaves,
  Where every joy prevailed,—
Fountains and flowers and laden trees,
  And love that never failed.

It chirped amid the buds of Spring—
  It drank the balmy ray—
And poured a more enchanting strain
  When woke the Summer day.

But when the leaves began to fall.
  At Autumn's frosty prime,
It seemed to meditate a flight
  To some more genial clime.

A flash of wings! too bright for earth—
  Wings from a cloudless sky,
Where happy clouds twine their wreaths
  Of immortality.

They came to lure it to a home
  Where Winter hath no sway—
And though the tears rained fast around,
  With them it soared away.

On the tenth day after the decease of Col. Colt, which took place on the 10th of January, 1862, his only surviving daughter, Henrietta Selden Colt, an infant of eight months, also died. The following lines were written by the Rev. D. E. P. Rogers, D. D., of New York, a relative by marriage of the Colt family, in memory of the twain so "lovely and pleasant to each other in their lives, and in their death not divided."

HENRIETTA SELDEN COLT.

Born May 23, 1861.—Died January 20, 1862.

Amid the growing forest
  There stood a noble oak,
In massive strength and grandeur,
  Unscathed by woodman's stroke,—
Rich, verdant Summer glories
  Encrowned its princely head,
While bravely to the tempest
  Its hundred arms were spread.

Beneath its sheltering branches
A lovely flower had sprung,
And with its infant tendrils
To the brave oak it clung;
It was a charming vision—
The flow'ret and the tree—
Of strength and beauty, blending
In graceful harmony.

There came a fearful tempest,
Careering in its might;
It smote the forest monarch
With its resistless blight—
Then drooped the tender flow'ret,
Left shelterless and lone,
Exhaled its dying fragrance,
And in a breath was gone.

And thus we learn the lesson,
Taught by our Father's care,
That neither strength nor beauty
The stroke of death may spare.
Thanks for the "better country,"
In endless life arrayed,
Where trees shall ever flourish,
And flowers shall never fade.

The largest slab in the family burial-ground is inscribed:—

SAMUEL COLT.

Born July 19, 1814.—Died January 10, 1862.

KINDEST HUSBAND, FATHER, FRIEND, ADIEU!

What visitors at this grave oftenest think of, regarding the sleeper there, may be, that his sun went down at noon. Yet, in contrast to the little ones around him, who died before they could say father or mother, how long his date of life, how far his wanderings over the earth, how manifold his experiences, how proud his achievements!

Interred in the midst of the garden his riches and taste created, and overlooking the industrial palaces his

15

genius evoked, he needs no encomium on his inventive-
ness and energy. Nor is his intellectual eminence in any
of its aspects what his friends most fondly recall. They
dwell rather on the moral traits to which the epitaph, in
the words, "kindest husband, father, friend," so touchingly
alludes. Indeed, it is an unmistakable token of the
wealth of his affections, that those who knew him most
loved him best, and will mourn for him longest. Nor
is there any one word in human speech more significant
than *adieu*, when rightly analyzed, as the commending of
a friend, for whom we can care no more and no longer,
to the hands of God.

> " Nay ! grieve not, being is not breath;
>   'Tis fated, friends must part,
> But God will bless, in life, in death,
>   The noble mind, the generous heart.
> So deeds be just, and words be true,
>   We need not fear stern nature's rule;
> The tomb, so dark to mortal view,
>   Is Heaven's own vestibule."

CABINET OF MEMORIALS.

ONE of the largest chambers in his residence appropriated by Col. Colt to festive occasions, has been since his death converted into a picture-gallery and repository of presents, arms, family portraits, memoranda, and scrapbooks of personal history. It is now (1866) on the eve of being still better adapted to this use, by the closing of its side windows and the opening of sky-lights. Light, hence falling direct from heaven, will set every object in the best point of view.

Chief among the curiosities here deposited is the Cabinet of Presents and Memorials. These are secured from meddling fingers behind a plate of glass stretching from floor to ceiling, and of proportionate width. The first received of all these valued gifts is a snuff-box of gold

from the Sultan, in 1850. On the sides are four medallions, each filled with tiny flowers of most exquisite enamelling; on its bottom, in one large medallion, lie some of Flora's loveliest children,—the rose, the lily, and the tulip: each smiles in its own fair hue, while the modest forget-me-not peeps out from among its more famed sisters. On its glittering top, arranged in floral forms, are as many diamonds as there are days in the year. These, set in silver on their pale blue ground, dazzle the eye and awaken the admiration of the beholder. When a classical scholar beholds such miracles of the goldsmith, filled only with a vile titillating powder, he is indignant that so noble a work should be designed for such a use. Nor can he fail of contrasting such a modern perversion with the casket of Darius, the most priceless of Persian spoils, set apart by Alexander to enshrine the Iliad of Homer, as the only gem worthy of such a setting.

To some fancies the Constantinopolitan box must have been devised for illustrating Hogarth's waving line of beauty, since it is everywhere curvilinear. Even the line of junction with its lid is a series of crescents. Yet it is not impossible that these moonlets were intended to bespeak its Turkish character.

One of the rings was the munificence of the Russian Grand Duke Alexander, in 1854. In the centre, on a dark-blue ground-work, is his imperial cipher surmounted by the crown, both in tiny diamonds, the whole surrounded by six brilliants of goodly size and the first water, while a fluted band of plain gold finishes the circlet.

Another and much more costly ring was at the same time the gift of his imperial father, the late Emperor Nicholas, Czar of the Russias. This has in the centre, on the same blue ground, the jewelled N and crown. Eight large diamonds surround this; the four largest being of uncommon brilliance, and seeming in their liquid beauty to be the home of the sunbeams. An almost innumerable number of small diamonds fill up the rare gift, leaving but a small portion of the plain gold visible.

The third ring, a testimonial from the late King of Sardinia, Charles Albert, was one of the earliest trophies of foreign appreciation of the revolver, and hence was bright as the morning star in the eyes of its receiver.

Its diamond, a superb brilliant, the peculiarity of which is to have an upper or principal octagonal face surround- ed by many facets, and which weighs considerably over seven carats, he had reset as an engagement-ring for the queen of his heart. This aristocratic stone,—a blaze of light,—well worthy of its Hebrew name, "stone of fire,"

shows costliness condensed, as when a vinaigrette of attar
corks up an acre of roses.   Regal   associations   also
heighten its lustre.   But to him who put it on the
finger of his betrothed, it was nothing more than it
would have seemed to Shakspeare, if he wrote the rap-
turous wooing-song which Stratfordian tradition ascribes
to him :—

> "Talk not of gems,—the orient list,
> The diamond, topaz, amethyst,
> The emerald bright, the ruby gay,—
> Talk of my gem, Anne Hathaway.
> She hath a way with her bright eye
> Their varied lustre to defy.
> The jewel she—the foil are they,
> My priceless pearl, Anne Hathaway.
> Anne Hathaway,—she hath a way,
> To shame all gems she hath a way."

All the rings are ponderous, as if to intimate that
potentates have things on a larger scale than other mor-
tals have hearts or purses for.   They recall those of
Roman prodigals, which Juvenal says must be burden-
some on a warm day ; and they justify the Roman jurist
who decreed that a blow from a hand with a ring on it
is twice as great a wrong as from one that has none.

Next and last received of the jewelled gifts is a golden
snuff-box from the present Emperor of Russia, Alexander Second.   The bottom and sides are of the most

richly chased and ornamented gold, but the cover is
shining with fair jewels. A medallion, more than two
inches in length, bearing the imperial cipher below the
jewelled crown, is set with thirty-eight fine diamonds; and
in the cipher and crown are seventy-two of varying sizes.
Outside this medallion and near each edge are six large
stones in solid crown setting, forming a most unique and
beautiful whole. Within, upon the highly burnished lid
of the box, is this inscription in Russe:—"From Alexander II., Emperor of all the Russias, to Col. Samuel
Colt."

There are also a set of vest and sleeve buttons, the tribute of a great-souled Texan. Meeting Col. Colt in Paris,
he praised the revolver, as over and over his life-preserver,
and begged him to accept some token of gratitude. His
gift, instead of the trifle that was looked for, turned out
a set of buttons, each of which was a jewelled star set
in gold.

Always and everywhere will jewels,—"the chief things
of the ancient mountains, and the precious things of the
lasting hills"—

> " Named in the foundation-stones of the bright city
> That is to be for the immortal saved—
> Their last and blest abode—"

fascinate all eyes ; but what a new brilliancy would be
theirs, did we believe in their mystical virtues! Could
we hold, as they did in the ages of faith, that diamonds
are sovereign against poison and madness,—that the tur-
quoise gives a start whenever peril is prepared for him
that wears it, and that the amethyst makes husbands
true,—then, without a metaphor should we speak of " ser-
mons in stones."

In the same group with the boxes and rings, imperial,
sultanic, and kingly, as well as the ranger's buttons, is
a Turkish order of nobility, which greeted him it was
intended to honor just before his last illness. This badge
bears Ottoman hieroglyphics, and is encircled by a series
of crescents, with each a star between its horns.

On the crimson cloth around these things rich and rare
lie twenty-four medals of honor.   But first in the mind of
him who received it was the large gold " Telford Medal,"
presented by the London Institute of Civil Engineers,
which society was incorporated in 1828, and before whom

he read a paper in 1851, and was thereupon elected their
first American member. The obverse is a bust of Telford;
the reverse, his masterpiece, the Menai Bridge, long the

most remarkable suspension bridge in existence. It bears
inscribed on its edge, COLONEL SAM. COLT, ASSOC. INST.
C. E. In the best portrait extant of Col. Colt, a full length,
16

of life size, recently completed by the great master, Elli-
ott, he stands holding in his hand this diploma, so highly
prized.

There are five smaller medals of gold from Connecticut
societies. Five from the American Institute, one as early
as 1845, "for the best guns, rifles, and pistols." In 1851
he received from the Exhibition of All Nations in London
a small bronze medal, of very fine workmanship, having
an excellent bust of His Royal Highness Prince Albert,
President of the Royal Commission, on its face. Also, in
1853, from the World's Fair at New York, a large silver
medal.

Another large silver medal, having on the obverse a
female figure just throwing open the doors of the palace
to Industry and Science, and on the reverse the outlines
of the building, with the words, "Crystal Palace, opened
MDCCCLIV."

Another large medal of silver, same year, bears on the
obverse the profile busts of Queen Victoria and Prince

Albert. The reverse is very handsomely ornamented with several female figures, and has inscribed upon it, "Ornatur Propriis Industria Donis."

From the French Exposition of 1855 there seem to have been three medals, two of bronze and one of silver.

On the face of one bronze medal is a striking bust of
Prince Napoleon, President of the Imperial Commission ;
the other bronze one has the busts of the Emperor and
Empress on its face, while the reverse shows an interior
section of the "Palace of Industry."

The silver medal is of very charming workmanship,
and the cut given here is a beautifully executed copy of
the reverse, the obverse showing a bust of "Napoleon III.,
Empereur."

After the coronation of the present Emperor Alexander
II., in 1856, at which time Col. Colt was attached to the
American Embassy at St. Petersburg, he received from
the Emperor a medal commemorative of his coronation, on
the one side the bust of the Emperor, surrounded by this
inscription in Russe :—" By the grace of God, Alexander II.,
Emperor and Autocrat of all the Russias." On the other,
done with superb skill, the Russian coat of arms, with
the double-headed eagle, the triple crowns, the globe, the
sceptre, the jewelled collar of the order of St. Andrew and

its heraldic devices. Above it the motto, also in Russe, "God is with us."

The same year he received another bronze medal from the London "Society for the Encouragement of Arts and Industry."

Perhaps the most costly of the medallic relics is a broad gold disk from the Sardinian King, Victor Emanuel II., in the autumn of 1860. This medal was brought to

Hartford by a committee of Italian gentlemen resident in New York, and presented to Col. Colt in person, with many expressions of gratitude for his substantial kindness to, and appreciation of, their honored countryman, Garibaldi.

The last medal he never saw, for his eyes were closed in death when it was received. It is of bronze, and very handsome. One face is given below. The other bears these words:—1862, Londoni, Honoris Causa—surrounded

by an exquisitely wrought wreath of oak leaves and
acorns.

Another shelf shows gifts from far Siam, where two
brothers rule an empire in harmony. They are called First
and Second Kings, both, however, being addressed as
"Your Majesty." The Second King, who has died during
this year (1866), was a man of considerable literary ability,
and wrote English exceedingly well, as his letters in
another part of this work show. The articles were all
made in Siam, and are of the best native workmanship.
The gift of the Second King, intended for table use, is
really beautiful, made of silver, but so heavily gilt as to
make one believe it to be gold, and is thickly embossed
with leaf-work and a sort of enamel.

From the First King there is a small snuff-box of pure

gold, a teapot and stand, a tea-caddy and cigar-case of the same workmanship as the table vase, but not as well finished.

The same capacious cabinet abounds in other notables, —Russian plate; ebony elephants; a Dutch mug found in an Indian's grave in 1858 or thereabouts, when building the Ferry Dyke, which was doubtless obtained from the Dutch at their settlement, scarce a mile above the spot where this little mug was buried by kindly hands, with the poor Indian, probably two hundred years ago; an anchor of California gold; a screen, the frame carved out of Charter Oak, and holding in ivory bass-reliefs the battles of Bunker Hill and New Orleans, which was the last gift to Mrs. Colt from her husband; a magnificent block of malachite; a Koran, with gold illuminations on every page; and some of the earliest forms of the incipient revolver.

In two side cabinets, which serve as pendents to the one now described, there are threescore and eight specimens of gun, pistol, blunderbuss, or wall-piece. The revolver appears in all its stages of development.

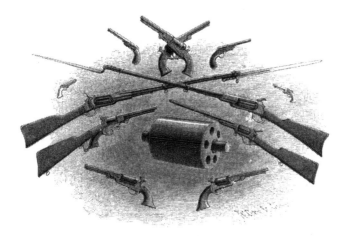

Among them is the so-called Texan model,—the variety which won its fame and fortune. Here is the identical weapon which the redoubtable ranger Walker carried at his saddle-bow all through the Mexican war.

17

There are multitudinous varieties of fantastic fire-arms. Some look long enough to touch most marks with their muzzles.

Nor are other arms here wanting. One dagger, from the East Indies, is serpentine in shape, as if it could thus steal the venom of the serpent, and is said to be poisoned. It was one like it which suggested this epigram on the belle dressing herself to the height of fashion :—

> " Adorning thee with so much art
> Is but a barbarous skill ;
> Like adding poison to a dart,
> Too apt before to kill."

Several claymores, from Japan, are without guards, and reputed to be of a steel no whit inferior to the blades of Damascus, which, as tradition says, were tempered in the coldest blast which in all the year howled down from Lebanon, where snow never melts. These are the self-same variety of weapon with which, in Japan, high officials, when guilty of foolish measures, are accustomed to do hari-kari, or execution, on themselves. Well were it for us, if it was only as easy to import the custom as it is the cutlasses!

Lowest in place, but not least in interest, in the miscellany, are some relics of the fire, which, on the fifth of February, 1864, consumed half the Arms-factory. All the fractions composing a revolver, thrown in a heap helter-skelter, may here be seen as they were welded into one mass, when the elements did melt with fervent heat, and the matchless order of the Armory—almost as suddenly as through the explosion of one of its builder's submarine magazines—sunk in a fire-chaos.

We have in another place described the costly and beautiful cradle made out of the wood of the Charter

Oak, which Hon. I. W. Stuart (the last proprietor of the old tree, in virtue of his being the husband of Miss Caroline Buckley, the accomplished daughter of Stephen Buckley, Esq., who purchased the estate of Wyllys Hill, on its sale by order of the Judge of Probate, for the heirs of the original settler, in 1823) presented to Mrs. Colt for her first-born child, in recognition of Col. Colt's interest in the preservation of the old tree, and the consecration of the Hill to public uses.

One of the choicest articles among the historical memorials at Armsmear is the Chair, which that ingenious artist, John H. Most, designed and wrought out of wood of the same old tree—in material, form, and style, worthy of a Roman senator in Rome's best days, or of Homer's

Jove in council of the gods—*dii majorum gentium.* The
following extracts from the published proceedings of the
Common Council of the City of Hartford, the letter of
Mr. Stuart, and the receipt of Mr. Most, will explain
the manner in which this precious article came into
the possession of Col. Colt.

"At a Court of Common Council of the City of Hartford, held on the 22d day of December, A. D. 1856—

"His Honor the Mayor having received a communication from Hon. I. W. Stuart, who wishes to present to
the Council fragments of the Charter Oak; on motion,
it was voted, That His Honor the Mayor, Messrs. J. B.
Crosby, and W. I. Goodsell, be a committee to receive the
same, and to cause a chair to be constructed from said
Charter Oak, for the use of the Council.

"Accepted and vote passed."

"At a Court of Common Council of the City of Hartford, held on the 12th day of October, A. D. 1857—

"The committee appointed to superintend the construction of the Charter Oak Chair, for the use of the Council, reported, verbally, that they could not approve of Mr. Most's bill for constructing the same, and wished to be discharged. Granted, and a new committee were appointed, who were authorized to examine the matter and report thereon."

"At a Court of Common Council of the City of Hartford, held on the 8th day of March, A. D. 1858, Alderman Merriman, Chairman of the Special Committee having the Charter Oak Chair under consideration, reported, 'That Mr. Most had made *three* propositions to the Committee. The first was, for the City to pay the whole amount of his bill, viz.: three hundred and seventy-six dollars. The second was, if the City, through the Council, would vote the Chair to the Connecticut Historical Society, he would reduce the bill to two hundred and fifty dollars. The third was, that Mr. Most would release the City from all claims he may have against the City, "providing the City released all claim to the said Chair." After due consideration, they recommend the passage of the original vote, as offered by Alderman Phelps, of January 26th, 1858, which is as follows, to wit: "Voted, That the City will release all claims to the wood of which the Mayor's Chair is made by Mr. Most: Provided, and upon condition, that Mr. Most releases the City for making the same."'

"Report was accepted and vote passed, eight to four."

"City Clerk's Office, ss. HARTFORD, *March* 12, 1858.

"The above and foregoing is a true copy from the

records of the Court of Common Council of the City of Hartford.

"Certified by HENRY FRANCIS, Clerk."

"HARTFORD, March 9, 1858.

"TO MY FRIEND, J. H. MOST:—I, I. W. Stuart, herein resign, set over, and deliver, all right, title, and interest whatever, in a chair known as the Charter Oak Chair, which the Common Council of the City of Hartford, in an unpatriotic fit of economy, as I deem it, has returned upon his hands.

" I. W. STUART."

"HARTFORD, March 12, 1858.

"The following is a true copy of John H. Most's receipt for all claims against the City of Hartford, for Charter Oak Chair, received and lodged with the City Clerk, March 11th, 1858, to wit:—

" 'HARTFORD, March 10, 1858.

"'I do hereby give up and surrender all claim upon the City of Hartford for the cost of manufacturing the Charter Oak Chair. And in conformity with a vote of the Common Council, take and retain the Chair, as my own property, to and for such use as I may choose.

" 'J. H. MOST.'

" Attest, I. W. STUART, E. SKINNER.

"Certified by HENRY FRANCIS, City Clerk."

"HARTFORD, March 10, 1865.

" Received of Col. Samuel Colt, four hundred dollars, in full for the famous Charter Oak Chair, made originally for the City of Hartford, for which it declines to make payment.

"JOHN H. MOST."

To this sum (four hundred dollars) Col. Colt added another one hundred dollars,—considering the "wood of which the chair was made more precious than cedar of Lebanon, and an article so beautifully wrought as this is cheap at any price." He had a profound reverence for such men as Wadsworth, the preservers as well as the founders of institutions by which man's essential inborn liberties are asserted. From his earliest boyhood he was the first to lead off in any public demonstration on the "glorious Fourth," and it was for what was thought, by college and academic authorities at Amherst, an over zealous and untimely display of patriotism on that day, "in the year of our Lord 1830, and of the Independence of the United States the fifty-fourth," that he left the academy abruptly and without permission, rather than make an apology for conduct which he thought "only mildly patriotic" towards the founders of the nation. To him, looking to the past and to the future of the great American Republic—

> "They were the Watchmen by an Empire's cradle,
> Whose youthful sinews show like Rome's; whose head
> Tempestuous rears the ice-incrusted cap,
> Sparkling with polar splendors, while her skirts
> Catch perfumes from the Isles; whose trident, yet,
> Must awe in either ocean; whose strong hand
> Freedom's immortal banner grasps, and waves
> Its spangled glories o'er the envying world."

Among the letters carefully preserved by Col. Colt was one addressed by Samuel Lawrence, of the firm of Lawrence & Stone, Boston, dated August 2d, 1830, inclosing the bill of ninety-one dollars and twenty-four cents for his outfit as a sailor before the mast in the good ship Corlo, Capt. Spalding, bound for Calcutta. Mr.

Lawrence adds:—"The ship sailed this morning. The
last time I saw Sam he was in tarpaulin, checked shirt,
and checked trousers, on the fore-topsail yard, loosing
the topsail. This was famous at first going off. He is
a manly fellow, and I have no doubt will do credit to
all concerned." It is not difficult to see ' the manly fellow,' with his brown locks curling out from beneath
his glossy tarpaulin, supporting himself with his feet on
the foot-rope, leaning on the yard, and, while loosing the
sail with his hands, casting happy glances at the crowd
on shore.

Mr. Lawrence, to enable the boy to bring home something from the far East, "told the supercargo to advance
him fifty dollars if he required"—little dreaming that
this same boy would bring back in his chest a little
wooden model, on which, during the voyage, he had
begun to whittle out the cylinder of the now world-

famous Revolver, which would revolutionize the construction of fire-arms throughout the world, and make a fortune for himself in a single year large enough to equip a fleet of merchant vessels for the farthest East.

The house where Col. Colt was born is standing, but changed in its outward aspects. The southeast chamber commands now the same outlook over city and valley as when it first met his eyes.

Among the epistolary curiosities here garnered are autographic letters from and to kings, dukes, and lords, as well as all other ranks of people. Some of them throw light on the progress of our mother tongue in the ends of the earth, or betray the mistakes into which those

18

who learn a foreign dialect more from books than from speech inevitably fall. The best of the whole collection, in point of chirography, are those of the Second King of Siam. Others were penned by men so extraordinary, that we would fain lose none of their utterances. Others are of growing value, as documentary evidence concerning some of the marvels from afar. All of them are important, as milestones in the progressive appreciation of repeating fire-arms. The note from Garibaldi has a special charm, having been written when he was about to launch forth, as most men judged, forlorn of hope, but, in truth, on the most triumphant scene of his almost fabulous career,—namely, to become the swift liberator of Italy, and to render her unity no longer impossible. It was prompted in acknowledgment of a donation of arms in aid of revolutionary Italy, and specially to equip the body-guard of Garibaldi himself. To the originator of repeating fire-arms, it must always have been a proud satisfaction, that he threw so many of them into the scale against the Neapolitan tyrant and in furtherance of Italian independence, and that at the crisis of crises, in the very opportunity of opportunity, when a feather might turn the trembling scale.

A selection from these letters will be printed at the close of this Memorial, chiefly in chronological order.

The Scrap-Books are full of the current history of his various enterprises, as these were chronicled in the press, or recorded in letters; and from one of these volumes we will construct a chapter on his early experience as an experimental lecturer, which we shall entitle the LECTURE EPISODE.

THE

LECTURING EPISODE.

THE LECTURING EPISODE.

In 1830, Samuel Colt shipped as a sailor-boy for Calcutta; in the year following he sailed back again. It is no wonder that he did not bring back with him the mines of Golconda. During several months afterward he was employed in his father's dye-works at Ware; but as the next spring opened, when not yet eighteen, and still "on the change 'twixt boy and man," he resolved, depending on his father no longer, in his own phrase, "to paddle his own canoe."

On the eve of Colt's setting forth upon his second adventure, his father bestowed his blessing in the form of the subjoined letter. Its touching words of cheer, of warning, and of love, went straight to the heart of the youthful wanderer, who carefully preserved it, and in after years indorsed upon it these words: "From C. Colt, Esquire, Hartford, March 30th, 1832. Advice on leaving home to seek my fortune."

LETTER FROM CHRISTOPHER COLT, ESQ.

My Dear Son:—You are once more on the move to seek your fortune, and must remember that your future prospects and welfare depend on your own exertions.

Should you seek diligently until you find some kind
of useful employ, I have no doubt but you may do well.
Do not despond, my son, but be resolute and go forward.
It matters but little what employ you embark in, if it is
but an honest one and well followed up, with a determi-
nation to excel in whatever you undertake. This will
enable you to obtain a good living, and to command
respect. Whether you go into a store, or go to sea,
or join in any kind of manufacturing, I deem it of but
little consequence, provided you devote yourself to your
employ with habits of close application; and all the leisure
time you can have, devote to study and sober medita-
tion, always looking to that kind Providence which gave
you existence, to be by Him directed in the path of
virtue and usefulness.

You will find it absolutely necessary to use the most
rigid economy. Until you have earned money and have
it in your pocket, never engage in any amusement or
unnecessary expenses.

When you get located, write me. In the mean time,
remember, I have more anxiety for your welfare than
can be expressed. I will write you my views when I
again hear from you. In haste,

Your affectionate father,

March 30, 1832. C. COLT.

"Do not despond, my son," said the fond father; but
how could the stripling help it? What could he do?
Ninety-one dollars and some cents had been all the
outfit that could be afforded him for India. Indeed,
Lawrence's letter indicates that he feared lest that sum
might appear extravagant to the elder Colt. Nor is it

likely that the youth was better equipped for his new
pilgrimage. No doubt the generous old man had said
to himself,—

> "Pity 'tis that wishing well had not a body in 't
> That might be felt. Then we, the poorer born,
> Whose baser stars do shut us up in wishes,
> Would follow with the effects of them our friends,
> And act what we can only think."

He deemed it indeed a pity; but it was no more pitiful
than true.

Many a looker-on would have judged young Colt to
lack internal capital no less than external. His school-
days had been cut short by a Fourth of July escapade in
Amherst. The romance of the ocean had faded away
during his two years before the mast. When a poor boy
says to a home-bound countryman, "May I ride on your
sled?" the answer is, "Yes, if you can get on;" and the
whip forthwith cracks up the horses. This is exactly the
usage which he who is setting out in life receives from
the world. Opportunities run from him at the top of
their speed.

But not even a tantalizing opportunity was at first
vouchsafed to Colt, so that he would have compared himself to the man who, coming home drunk, gropes at his
door in the dark, till, finding no aperture for his night-
key, he begins to swear that some thief has scampered
off with the very keyhole.

What could the stripling do? He could do every
thing, if the pistol and pyrotechnic ideas, already seething
in his brains, could only be realized and utilized by
experiments. But for such tentative processes money,
"which answereth all things," was the one thing needful.
"Give me a stand-point," said Archimedes, "and I will

move the world." It was such a πoῦ στῶ, or base of operations, which Colt made his first object, and in securing
which, in the teeth of all discouragements, he first manifested his youthful genius.

What could the stripling do? Within three months
of the father's farewell epistle, the following editorial notice was issued in the Boston *Morning Post* for Saturday,
June 22d, 1832 :—

"The exhibition of the singular and amusing effects of
the nitrous oxide gas, when inhaled into the lungs, will
be repeated this evening only, at the Masonic Temple.
(See advertisement.)"

The hero of this phenomenon was no other than
Samuel Colt, under the incognito of Dr. Coult, a sobriquet
which he continued to wear for several years thereafter.
Already had the stripling won a position among lecturers,
—the Italian improvisatori of New England.

However his education had been in some points neglected, it is not to be denied that he had, from first to
last, made the most of all his advantages for studying
chemistry. Not •only had he become an adept in the
various processes of the dyeing and bleaching factories,
but he had spent a winter in the laboratory of a chemical enthusiast, and thus was no mean proficient in many
an adroit and startling manipulation. In this career he
had instinctively followed a time-honored maxim, which
possibly he had never read :—

> "Study what you most affect.
> No profit grows where is no pleasure ta'en.
> Small have continual plodders ever won
> Save base authority from others' books."

Thus it came to pass, that becoming a chemical lecturer called into full play his best knowledge, and brought out his strongest point. His choice of such a career proved that he knew himself—knew his "forte," —and self-knowledge, although no science, is fairly worth the seven.

The time, too, was auspicious for the new enterprise. In those days lecturing was just beginning to popularize the results of science and learning, and had not as yet lost the gloss of novelty. On the contrary, as an auctioneer of knowledge calling all to buy a pennyworth,— as an itinerant college carrying the mountain to Mahomet when Mahomet would not come to the mountain,—it was believed about to work greater miracles than are any longer expected from such a source.

Chemistry, moreover, as a new science, was welcomed by the masses with the bloom of young desire. Experiments have also, always and everywhere, an intrinsic charm for auditories, since their eyes are more learned than their ears, so that the former catch in an instant what the latter cannot learn in an hour. Nor is there a chemical marvel more available for popular impression than the exhilarating properties of nitrous oxide gas when inhaled.

To discover them, about thirty years previous, had made the fortune of Sir Humphrey Davy, a stripling not out of his teens, and who had always before been stigmatized as "an idle, incorrigible boy—of no taste for the classics." To display these properties of vital oxide in a hundred towns where a chemical lecture had never been heard of—to invite every man who wished it, to breathe what the poet Southey had said must compose the atmosphere of the very highest of

19

all heavens—to expose before each assemblage the more venturesome of its members harmlessly making fools of themselves, dancing, boxing, fencing, laughingly or lack-adaisically, with the curtains of all ceremony rolled up to the clouds—to let loose each ruling passion in a sally of transient abandon—to pour out for men the hilarities of intoxication without its headaches—to do all this could not prove a thankless or a gainless task. What pro-gramme could be more fascinating? It had proved so even in the lecture-room of Silliman and Hare.

The Esquimaux live on train-oil; yet they dare not take it clear, as it would be too hearty food. So they mix it with sawdust. On the same principle, all men, however grave, demand a sauce of humor to help them digest the strong meat of wisdom with better appetite, and he who peppers the highest is surest to please.

Hence Sheridan long ago said, "No speech can fail to be a bore if it lacks a jest." Every lecture which aspires to be popular must *pop*. Humor is the only sop which can make the many-headed barking Cerberus of hearers fawn upon a speaker. A Governor, Senator, or Major-General may be asked to speak once in virtue of his office, but even men of the most grim and vinegar aspect will never go to hear him a second time, unless he makes them perforce show their teeth by way of smile. The history of all unmirthful lecturing, except now and then a Caudle lecture, is written in the doggerel—

> "All Greta's snobs,
> By sundry jobs,
> Were drawn to hear a lecture;
> But soon forgot
> The lesson taught,
> And yawned beyond conjecture."

The more mouths open to swallow a speaker, the more minds are shut against his speech.

The secret of producing a lecture duller, if possible, than preaching, is to leave out all spice of humor and all salt of wit. But who in our age, Gough only excepted, can boast such a pregnant wit, set off by such mobility of form and feature, that ño man of the millions who look on his countenance can keep his own? Now all the comic feats which the man of all men—at whose birth the Muses not only smiled, but all laughed outright—can achieve, the imbibers of laughing-gas were ready to pay the nitrous oxide lecturer for liberty to perform. He put them into nets and cages, but when they felt a fever of his madness they uproariously broke out—and each of their tantrums served him as a gratuitous advertisement. The Reineke Fuchs of Kaulbach is unrivalled for portraying developments of human nature through animal forms; Colt, in his volunteer menageries, thanks to the witchcraft of gas, was equally successful through presenting animal natures as betrayed in such as wear the forms of men, when forced to drop the mask. Hence, while other showmen must carry clowns about with them, he found a plenty everywhere ready to his hand.

The lecturing tours of Dr. Coult may be traced through a period of well-nigh three years,—and in all the length and breadth between Quebec and Charleston, New Brunswick and New Orleans. This career, which his favorite study, his noble form and features, as well as the temper of the times, combined to render reputable, gainful, and congenial, taught him as much of the land as his India voyage had revealed of the water. It was not only what

Mrs. Sigourney used to call her pen,—his bread-winner,—but it yielded him material aid for brooding upon and hatching his inventions. Nor, to a spirit like his, are such travels and leisurely surveys any thing less than a Liberal Education.

The dates of his visits in seventeen widely severed places are stated in the contemporary local papers as follows: In 1832, June 22d, Boston. No trace of him appears for more than a year, but he was probably electrifying the West, far and wide. In 1833, on the 7th of August, he reappears in Virginia, at Wheeling. On August 17th he was in Pittsburg; on the 19th of September, in Rochester; in Albany from the 12th of October till the close of the month. Again he vanishes for three months; but in 1834, on the 1st of February, turns up at Winchester, Virginia. On the 4th of April he was in Charleston, and on the 18th in Augusta. On the last day of June he lectured in Albany; on the 17th of July, in Montreal; on the 2d of August, in Quebec; on the last of September, in St. Johns; in October, in St. Andrews and Fredericton; then at Portsmouth and Newburyport, and at Lowell on the 29th of November. In 1835 we have no documentary evidence that he lectured anywhere save in Petersburg, Virginia, on the 17th of February.

From a comparison of these dates, it is obvious that he between whiles gathered audiences in a multitude of other places. The editorial notices from which the above details have been culled are significant. They make it clear that the lecturing doctor, not content with formal advertisements, left no stone unturned to fill his houses by the aid of that editorial small talk, *in leaded types*, which is all in all with most readers, and the real manufacturer of public opinion. Indeed, lecturing conquests have seldom. been pushed so fast and far as his without a good deal of tact as well as talent—not to say *gasing* and gasconading. Hence it is not incredible that, in the outset, he parted with his last dime to procure so small a crumb of comfort as the nonchalant item above cited from the Boston *Post*, and that, perhaps, after emptying his purse for gaining permission to make his *début* in the Masonic Temple. He knew that to be spoken of as a Lecturer at the Hub of the Universe, and in its most aristocratic hall, would enable him to radiate on any spoke he pleased to its very circumference. The key-note of his policy in lecturing, as in every other sphere, sounds forth from his attitude when last seen, as he started for farthest Ind, "standing on the fore-topsail yard, and loosing the topsail."

In some of the "hurry-graphs," as Willis terms them, reporting or heralding his performances, there is praise of scientific amusement mingling the *utile* with the *dulce;* in one he is described, after inhaling his "airs from heaven," as singing " *The Merry Swiss Boy;*" others are comments, sometimes apologetical, sometimes derisive, on the fantastic tricks of the gas-bibbers; another proclaims that the gas was to be imbibed by six Indians; others characterize the lecturer as a medical gentleman, an eminent chemist, a practical chemist, etc., from New York. In the far Southwest they might have added, *and a good long way from it, too.*

At least one of the hand-bills, and one reporter's sketch, deserve preservation here, as marking an interesting phase in a many-colored life.

SCIENTIFIC AMUSEMENT—NITROUS OXIDE GAS.

Dr. Coult (late of New York) respectfully informs the Ladies and Gentlemen of Lowell and vicinity, that he will Lecture and Administer Nitrous Oxide, or exhilarating Gas, this evening, Saturday, Nov. 29, at the Town Hall. The peculiar effect of this singular compound upon the animal system was first noticed by the celebrated English Chemist, Sir Humphrey Davy. He observed that when inhaled into the Lungs, it produced the most astonishing effects on the Nervous System; that some individuals were disposed to Laugh, Sing, and Dance, others to Recitation and Declamation, and that the greater number had an irresistible propensity to muscular exertion, such as wrestling, boxing, &c., with innumerable fantastic feats. In short, the sensations produced by it are highly pleasurable, and are not followed by debility.

Dr. C. being a practical Chemist, no fears need be entertained of inhaling an impure Gas; and he is willing to submit his preparation to the inspection of any Scientific Gentlemen.

Dr. C. has exhibited the extraordinary powers of this Gas in many cities in America, to numerous audiences of Ladies and Gentlemen of the first respectability. He has administered it to more than 20,000 individuals, and has taken it himself no less than 1,000 times.

The persons who inhale the Gas will be separated from the audience by means of a net-work, in order to give all a better opportunity of seeing the exhibition. Dr. C. will first inhale the Gas himself, and then administer it to those who are desirous of inhaling it.

Such individuals as wish to inhale the Gas in private parties, will be accommodated by applying to Dr. C. at the Merrimack House.

Tickets 25 cents each, to be had at the principal Hotels and at the door. Seats may be secured between the hours of 12 and 3 o'clock. Doors open at 6½ o'clock; entertainment will commence precisely at 7.

AMUSEMENT SCIENTIFIQUE.—Gaz Oxide Nitreux, Encore Trois Soirées, à la Chambre d'Encan de M. Cole, près de l'Eglise Anglaise. Le Dr. S. Coult informe respectueusement les Dames et Messieurs de Quebec et de ses environs, qu'il discourra sur le Gaz Oxide Nitreux ou Vital, et administrera ce Gaz Lundi, Mardi, et Mercrédi Soir, le 4, 5, et 6 Août, 1834.

L'effet particulier de ce Gaz singulier sur le système animal, a été remarqué pour la première fois par le célèbre Chimiste, Sir Humphrey Davy. Il observait que lorsque ce Gaz était introduit dans les poumons, il produisait les effets les plus étonnants sur le système nerveux; que quelques individus étaient disposés à rire, à chanter, et à danser, d'autres à réciter et déclamer, et que le plus grand nombre éprouvaient un penchant irrésistible à faire usage de leurs facultés physiques, en donnant des coups de poing, et faisant un nombre innombrable de mouvements fantastiques. Enfin, les sensations produites par ce Gaz sont délicieuses, et ne sont point suivies de débilité.

Le Dr. C. étant un Chimiste pratique, on ne doit point appréhender de respirer un Gaz impur; il soumettra volontiers ses préparations à l'examen des savants.

Le Dr. C. a exhibé le pouvoir extraordinaire du Gaz dans plusieurs Villes de l'Amérique, devant des assemblées nombreuses de Dames et de Messieurs de la plus haute distinction. Il l'a administré à plus de 20,000 personnes, et l'a pris lui-même plus de 1,000 fois.

Les personnes qui prendront le Gaz seront séparées de l'auditoire au moyen d'un réseau, afin de faciliter à tous le moyen de voir l'Exhibition.

Toutes les peines possibles ont été prises pour que la maison soit bien aérée, et qu'on s'y trouve à l'aise.

Toutes les personnes qui désireraient prendre le Gaz dans des réunions privées, pourront s'adresser à cet effet au Dr. C. à l'Albion Hotel.

Billets d'Entrée—Sièges de Devant, 2s. 6d. Sièges de Derrière, 1s. 3d. On peut se les procurer à l'Albion Hotel, et à la Porte. Les Places peuvent être retenues depuis Midi jusqu'à 2 heures. Les Portes s'ouvriront à 8 heures, et l'Exhibition commencera à 8¼.

Quebec, 2 Août, 1834.

From the Albany Microscope, October 26, 1833.

MUSEUM.

We never beheld such an anxiety as there has been during the past week, to witness the astonishing effects of Dr. Coult's gas. The museum was crowded to excess every evening; and so intense the interest which was manifested, that the doctor has been compelled to give two exhibitions almost every evening.

The effect which the gas produces upon the system is truly astonishing. The person who inhales it becomes completely insensible, and remains in that state for about the space of three minutes, when his senses become restored, and he sneaks off with as much shame as if he had been guilty of some little mean action. ☞ No per-

son will begrudge his two shillings for the gratification
of half an hour's laugh at the ludicrous feats displayed
in the lecture-room.

From the Montreal Vindicator, July 18, 1834.

NITROUS OXIDE GAS.

Dr. Coult's Exhibition presents some of the most pleasing and laughable scenes one can well imagine. Although
the peculiar effects of Nitrous Oxide keep the audience
in a state of almost continual merriment, yet there is a
great chance for the learned and curious to exhaust all
their wits in sober contemplation on the causes and
effects of Nitrous Oxide upon the human system.

It seldom happens that any thing like Scientific Amusement comes within the reach of our citizens, and consequently all should give it their encouragement. Not the
least shade of impropriety attends the amusement. An
evening cannot be spent more profitably and pleasantly
than at Nelson's Hall.

On the whole, if much of the doctor's lecturing was
rather superficial than scientific, and not the best he could
give, it was still as good as the bulk of his hearers
would receive. It was at least nonsense suited to the
popular nonsense, and hit the public taste with a broad-
side between wind and water. Accordingly, it was well
that our hero assumed the title of Doctor, since his spec-
tacles provoked laughter so side-splitting as often to call
for surgical aid.

Now and then, however, it was a little embarrassing to
the lecturer that he was reputed a Doctor. Once, for
instance, as he was steaming up from New Orleans, the
cholera broke out on board the boat,—for almost the first

20

time on western waters. Nor was there any other physician among all the passengers. His services were of course demanded in a way which admitted of no refusal or evasion. It was of no use to talk about an unheard of disease, or of lacking medicines. So the doctor set to work,—administered gas and chemicals which he knew could do no harm, and through such placeboes so acted on imagination that there were some surprising recoveries, among them that of one of those famous singers, the Ravels.

But Col. Colt's lecturing and change of name form only a brief and sportive episode in his biography. They evince his versatility, but are mainly memorable as a specimen of that lowliness which is young ambition's ladder. They were resorted to as is that temporary scaffolding which assists in the rearing of a palace, or as school-keeping is the by-business of students who are working their passage through college. In this view, his portable laboratory, which, as Silliman used to say concerning that in Yale, when he first came there, was no larger than a candle-box, was not a few retorts and nitrates, or gaseous bases, but the potentiality of progress and riches beyond the dreams of conservative avarice. Besides his own testimony on this point, often repeated in conversation, and written out in his letter to President Tyler, we have the following cordial epistle,—such as those only can write who feel that old friends can no more be extemporized than old trees,—from the sculptor, Hiram Powers.

FLORENCE, *Sept.* 10, 1851.

MY DEAR FRIEND COLT:—I have at last got possession of the wonderful "revolver," which you have been so very kind as to send me.

And now let me thank you for this token of your kind remembrance. Our experience has been somewhat similar. Both of us have had some tough times in our day, have passed through a variety of trying processes upward to some distinction. But you have found pecuniary reward the soonest, and well have you merited it, and as well, the high reputation you now enjoy.

I shall never forget the *gas* at the old Museum (in Cincinnati), nor your sly glances at the ropes stretched around the columns, when about to snatch the gas-bag away from the huge blacksmith who glowered at you so threateningly while his steam was getting up,—nor, a moment after, his grab at your coat-tail, when you, frog-like, leaped between the ropes.

I remember your telling me in Washington, that at that very time you were elaborating in your mind the great invention you have since given to the world. But little did I then dream that in 1851 I should be in Italy, a sculptor, and fully employed. I had hopes of better things for the future, but they were faint indeed.

I have not forgotten my old mechanical pursuits, [he alludes to various contrivances in Cincinnati, the chief of which, styled the Infernal Regions, horrified thousands of visitors at the Museum,] and I have a shop even here,— a turning-lathe, forge, &c., and I spend much of my leisure time in this way. I have invented several improvements in working marble and plaster of Paris. One of them you shall see one of these days, for it embraces much of your own art. It could not fail to be very useful even to you. It is not yet complete, but will be in a few months.

Jonathan is indeed taking a stand among the nations

of the 'arth.' If his show at the Great Exhibition is meagre, he nevertheless beats all creation in his threshing-
machines, his steamers, and yachts. John Bull don't like
this from his rebellious son, but he chuckles at his being
HIS son, after all.

Wishing you increased success in all your undertakings,
I am, my dear friend,
Ever most sincerely yours,
HIRAM POWERS.

It is by no means incredible that it was in some exhibition got up by Messrs. Powers and Coult, that when
an accidental explosion of chemicals blew up the spectators as if a torpedo had ignited under them, a sailor, who
had been amazed at a good many of their previous
tricks, thought this catastrophe also only a part of the
show, and when he began to recover his senses, cried
out, with an oath, to a comrade, "I wonder, Jack, what
in the world they'll do next!"

His lecturing purposes once accomplished, Samuel
Colt emerged from his incognito and itinerant shows as
eagerly as from purgatory. The same year he crossed the
Atlantic in quest of European fire-arms patents, and the
very next year obtained one in Washington; henceforth
soaring cloudward, but not scorning the base degrees by
which he did ascend.

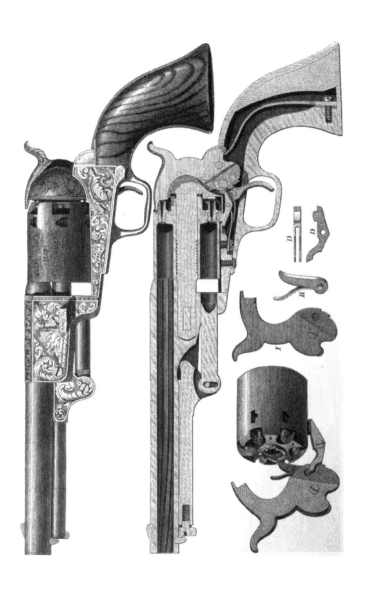

THE

COLT REVOLVER.

THE

OLT REVOLVER.

THE COLT REVOLVER.

SAMUEL COLT'S was a brain teeming with shrewd inventions, but the one which from first to last engrossed most of his time and enthusiasm, the corner-stone of his fame and fortune, was his pistol. Herein his great strength lieth. In him, as in the Olympic race-course, many competitors ran, but one only received the prize.

What did he do to improve the pistol? In one word, as Augustus found Rome brick and left it marble, so Colt found the pistol a single-shooter, and left it a six-shooter. Thus judged the Texan rangers, when they coined this new word "six-shooter," to describe a thing no less new among men, an engine which rendered them victorious against fearful odds, and over both Mexicans and Indians.

In more than one country, and in more than one past age, fire-arms had, indeed, been fabricated, each with a revolving cylinder containing several chambers, intended to be loaded at once, and then to discharge all their balls through a single barrel. But all these contrivances, however admirable as "tricks to show the extent of human brain," had proved abortions, visionary impracticabilities, and devoid of all practical value. The lack of percussion-powder as priming, the use of which in the

British army was first adopted as recently as 1840, would
alone have sufficed to make their success, if not impossible, at least very partial and imperfect.

None of these repeating shooters could boast any
efficient expedient for turning the cylinder in the act
of cocking a gun. To guard against the simultaneous
explosion of more than one of the charges, all of them
needed a cover over each priming-pan.

What they, one and all, promised, was the heart's
desire of every military man—and hence, had they not
labored under fatal defects, they could not, age after age,
have remained mere curiosities. They could not have
failed to give birth to many a factory, as Colt's idea has
done—making the world aware that he was in it. Such
an idea, once developed so far that it can be utilized, cannot be hid, more than a city that is set on a hill. It
speaks to men as potently, as the first lump of silver
unearthed at Potosi, called them to dig there for other
nuggets of like lustre.

Samuel Colt, at a time when, according to his own
testimony and all evidence, he had never seen any of these
strange devices for multiplied discharges without reload-
ing, invented and constructed an arm which accomplished
the consummation that had been the end and aim of so
many baffled endeavors. Henceforth, every man armed
with one of his weapons was armed sixfold. One pioneer
on the Merrimac, when assailed by a party of Indians,
through pointing his gun now at one and then at another,
but withholding his fire, escaped their knives himself, and
saved his seven children. If his piece had been a revolver,
he might also have shot five of those miscreants. Nay,
had his hand borne a six-shooter magazine, the savages

might not have attacked him at all, for they would have seen that though they drew his fire, five times over, the first man of them who rushed forward with the tomahawk would rush on certain death. A six-shooter is even more than six times a single-barrelled pistol, because it bears their effectiveness all condensed and massed in its single self, just as a fist is something more than all the fingers and the thumb which compose it. While the progression of increase in its barrels is arithmetical, that of its potency is geometrical. Spite of mathematics, it is such an integer that its total impression is more than that of all its component fractions. Force so concentrated and held in reserve for the opportunity of opportunity cannot but "demoralize" antagonists. Each fresh chamber, as it sweeps round into line with the barrel and hammer, resembles a new head of the multiplying hydra, which, rising in place of the old one his club had crushed, thunderstruck and daunted Hercules himself.

Fitly, so far as we may compare small things with great, we may apply to it Canning's simile concerning England and a ship of the line, "which, whilst apparently passive and inert, silently concentrates power to be put forth on an adequate occasion, when springing from inaction to the display of its might." When the artist Catlin, exhibiting a revolver in an Indian tribe where one had never been seen, had planted five or six balls in a target, and made as though he would fire more, the Chief said to him, "Stop—I am satisfied you can fire all day without loading, and it is a pity to waste ammunition." He would naturally have named such a wonder-working machine "the pistol-God." Ovid would have described it as the thunder-

21

bolt of Jupiter, omnipotent as ever, yet sure to be stolen
by Cupid, because no longer too heavy for him to lift, or
too hot for him to handle; the thunder of Jove in the
thimble of Minerva! *Venus armata.*

The first "pistol idea" or original model struck out
by Col. Colt, in 1830, when at the age of sixteen he
was a sailor-boy before the mast, on a voyage to India,
consisted in "combining a number of long barrels so
as to rotate upon a spindle by the act of cocking the
lock." This many-barrelled contrivance, long after it had
been discarded by Colt, was in part borrowed in one
variety of pistols, familiarly termed "Pepper-boxes,"
by those who feared to infringe his patent, yet strove
to steal some of his profits. But through fear of the
patent some of the pepper-boxes dared not revolve,
but were fired by a revolving hammer. Such an unwieldy arm, of whatever pattern, contrasts with the real
revolver as much as the rooster the Mohammedan pilgrim to Mecca carries in his wallet, to wake him up for
nocturnal prayer, is more cumbrous than an alarm-watch
or repeater.

To the quick wit of young Colt it soon appeared
better, for lessening weight and bulk, to have only a
rotating cylinder containing several chambers, and to
discharge all of them through one single barrel. For
this device he obtained a patent in 1836, and this implement was a six-chambered but single-barrelled weapon
which could be *used*,—and the first not revolving by
hand ever made which could be,—indeed, which needed
but little more care to keep it in order than one of
a single barrel.

The schedule which Col. Colt drew up, describing

the parts and combinations, in his first application for a patent, is no adequate description of his arm as now manufactured. It was still complicated and clumsy, for the same reason which a certain minister assigned by way of excuse for preaching a long sermon, namely, that he had not yet had time enough to make it short. Nothing but time enables to distinguish the essential from the accidental, and so to ascertain what it will do to retrench.

After all, the principle of the invention lies in that first pistol, however defective in mechanical execution, and whatever after-thoughts accessory to its effectiveness it lacked. Many a better vessel than the bark of Columbus has crossed the Atlantic. Nevertheless, it was his ship and not another's,—his ship, however poor, not any clipper, however fleet, nor the *Great Eastern*, however stately,—which discovered the new world.

But let us suppose that in Herculaneum, or in some other out of the way place, an arm should be raked out of the dust of old oblivion, by some mousing antiquarian, which revolves in the act of cocking—what shall we say to it? Why, that it militates against the independence and utility of Colt's invention no more than the discovery of America by Scandinavians, who voyaged hither five hundred years before Columbus, but never told anybody where they had been,—a fact now long admitted by all scholars,—will detract from the glory that crowns Columbus as an independent discoverer of the new world, and, for all practical purposes, the original and only one.

It is a curious fact, that more than a dozen years

ago a fire-arm identical in principle with Colt's was
said to be discovered. It was brought into court in
Boston, by parties whom he had prosecuted for trespass-
ing on his patent, as proof that he had originated
nothing new. Its mechanical execution was as coarse
as that of orientals, and it appeared centuries old. It
was at first a poser to Colt's counsel. But no sooner
did one of his workmen get permission to take it in
pieces, than it betrayed indubitable marks of modern
origin, and was seen to be a Yankee copy of Colt's,
clumsily fabricated on purpose that it might pass for
an antique. It thus came to resemble a gun which,
bursting in the act of discharge, hurts him who fires
it more than him it is aimed at. Nay, it did much
to decide the suit against its owners. Court, jury, and
all candid men felt that had any genuine gun existed
involving Colt's principle, his opponents would have
produced it, instead of forging a counterfeit.

The first patent granted in this country to Col. Colt
was issued by the authority of Andrew Jackson, and
bears the date of the twenty-fifth of February, 1836,
although a patent had been taken out in England in
1835. For a quarter of a century thereafter his cardinal
idea, at home and abroad, amid evil report and good
report, through success and through failure, making use
of every friend and no less of every foe, like Bacon, glad
to light his torch at any man's candle, in poverty and in
opulence, was to heighten by every subsidiary prop the
perfection of his six-shooter. This was his initial and
also his terminal point.

"I honor Columbus," said Turgot, "not because he
discovered America, but because he sailed in quest of

it through faith in an idea." Herein also lies the merit of Colt. He had such faith in his convictions, that, like Columbus, he acted on them all through his life. He made such manifold experiments on his favorite implements, that he could have equipped a company of soldiers with weapons he had thrown away as of impracticable patterns, and yet have given each man a fire-arm different from all the others. His success, then, was well earned, for it was the last result of countless failures—failures amid which the patience of other inventors in that line had succumbed. Such is the price of success. Whoever is successful—in distinction from him who is merely fortunate, thanks to some lucky blunder—has persevered in spite of multitudinous failures. When George the Fourth admired the inimitable yet simple grace of Beau Brummel's neck-tie, *simplex munditiis*, that autocrat of fashion remarked, "Your Majesty little knows how much that trifle, which appears finished more through happiness than pains, has cost me."—"Why," said the king, "not many shillings, at most." Thereupon the dressy genius and magnificent trifler, opening a door, showed the sovereign, in his toilet-room, a heap high as a haycock of neck-cloths which he had spoiled in vain endeavors to snatch a grace beyond the reach of art, and beyond the cavils of fastidious critics, of whom he was himself chief. The perfection of art is to hide all art.

What was the pistol when Colt had laid his last hand on it? It was as perfect in its way as Beau Brummel's neck-tie, for it had cost proportionably as much, and it bespeaks in its improver a character as much nobler as an earnest investigator is above a professional dandy, and as an invention that decides the destinies of nations, is superior to a prettiness which amuses fashionable exquisites. What is it? It is a *novum organum*, "a new tool" not unworthy to be named with the masterpiece of Bacon. It is a miracle of much in little. It is a six-shooter, the largest size of which for the pocket, or police, with a barrel half a foot long, weighs four ounces less than two pounds.

In it power and portability rejoice together. You can fire it more than a thousand times without cleaning. No matter if the barrel becomes foul, a ball will go through it, though wire-drawn into a knitting-needle in its passage. Strike the hammer of an old-fashioned percussion pistol and you fire it off, if it be loaded. You cannot thus explode Colt's, for its hammer rests, not on a cap, but on a safety-pin between cap and cap. As you are cocking a Colt pistol, its cylinder revolves just far enough to bring a cap under the hammer, no more and no less. Thus, it cannot turn too far, nor at the wrong time, as when full cocked, which would prevent the hammer's hitting the cap, and the ball's passing into the barrel. Nor, on account of the safety-pin, can it turn when uncocked, which would bring the hammer over a cap, as in the old style of pistol, so that a blow on the hammer, or raising it a little through any accident, would occasion a discharge. Draw now the hammer to half-cock, and the cylinder revolves freely, so that you can bring its chambers one by one beneath the ramrod for loading. Many a soldier in action has lost his life while loading, simply through the time lost by dropping his rammer. Colt's cannot be dropped, being hinged on the body of the weapon. Other ramrods will not work after the barrel becomes foul. This never enters the barrel at all, and, acting as a lever, more than quadruples the power of the loader's arm, so that it can with ease thrust into the chambers balls which fit water-tight. Throw one of Colt's weapons, with a bit of wax on the nipples, into water, and when you take it out it will not hang fire, for the cap at one end and the ball at the other hermetically seal the cartridge. Captain S. H. Tobin, a Texas Ranger, declares that he saw a Colt pistol, which had lain, while loaded, exposed on the prairie to the storms of three months, when fresh caps were put on the nipples, go off as well as if charged only the day before.

Had Colt's cartridges come into vogue a few years sooner, it is among possibilities that they would have averted a rebellion second in its extent and atrocities only to our own,—that of the Sepoys in India. The chief cause of that wide-wasting civil war is stated by high authorities in these words: "The government had resolved to arm the Sepoys with Enfield rifles, and a new kind of cartridge, greased in order to adapt it to their bore, was accordingly introduced in the depot of musket instruction at Dumdum. A report spread among the native troops that the English intended to make them give up their religion, by causing them, as the cartridges in loading had to be torn with the teeth, to bite the fat of pigs and of cows, the former of which would be defilement to a Mussulman, and the latter would be sacrilege in the eyes of a Hindoo." More than by any other single grievance, were the native soldiers infuriated to mutiny and rage through fear of pollution from animals they abhorred as accursed, or of

becoming worse than cannibals through eating of those
they revered as divine. But Colt's cartridges, need-
ing no tearing, could not raise scruples in any Sepoy, though of the straitest sect, whether Moslem or
Buddhist.

All old-fashioned hammers struck downwards; hence
their jar tended to disturb steadiness of aim; while Colt's,
striking exactly in the line of your aim, have no such
tendency. In his arm, again, the percussion sparks, being
poured as through a funnel, concentre on the charge,
while according to the old arrangements they were
scattered from a focus. This old fashion of inserting
cones wrong end foremost—now that Colt has taught
the world better—has come to appear no less absurd
than General Pillow's order in the Mexican war—that
an embankment be thrown up on the wrong side of
a ditch. Gun-makers, however, were far from discover-
ing their blunder so soon as our general who intrenched
himself only against a fire in the rear, but whose eyes
were opened by the first discharge from his enemies'
batteries.

All revolvers before Colt's were destitute of a priming apparatus, which, while not slow and awkward,
would be safe against the simultaneous discharge of
contiguous chambers. His was a masterpiece because
it overcame this difficulty, not only by employing caps,
and that before they were allowed in any army, but
by projections between cap and cap. Lest the cylinder
and barrel should not be kept in close contact, he
joined them together by a key. While it was of
vital importance for guarding against a lateral discharge
that these pieces, like Shakspeare's friends, should be

grappled to each other by hooks of steel, they were in danger of being wedged together so tight that the cylinder would not turn. Hence, Colt checked the key with a tempering-screw.

All these movements and checks,—yes, and too many others to be enumerated,—each seasoned by seasonableness, never in the way and never out of the way, are effected by the combination and interaction of half a dozen bits of iron, and that in a nut-shell — operations eccentric, intervolved, yet regular—then most when most irregular they seem.

Among many experiments tried by Col. Colt, one was to both cock a piece and fire it by the same drawing of the trigger. But with arms formed on this principle, it was found impossible to take and to keep a steady aim. Another was a ring trigger beneath the lock, to raise the hammer and rotate the cylinder; but in this way the complexity of the lock was doubled. Another was a lubricator, consisting in a sort of ramrod gliding down into the chambers of the cylinder one by one, and at each descent leaving a drop of oil. This device, also, if sooner projected, might likewise have warded off the Sepoy ·rebellion in the East. Yet, for the new style of cartridges, it is as needless as the fifth wheel of a coach. An additional variety of lubricators was fitted into the carbine breech, when made so as to be attachable to the holster-pistol. This was a pint flask, or canteen, which would hold water, and probably whiskey, if the trooper could get more than he wished to drink on the spot. This last species of lubricator was voted worthy of all acceptation alike by officers and privates, whatever their views on temperance. Although captains

22

complain of their men that they lose many arms, it was confidently predicted that while the carbine might be missing, the carbine attachment never would be. It bids fair, accordingly, in some form, to remain attached to every army corps to the end of all wars, which is as much as can be expected for the revolver itself. Some admirers of Falstaff will call him the original inventor of the patent lubricator, since in one of his holsters he substituted a bottle of sack for a pistol.

For a time Colt's arm was manufactured with a groove cut in and around the base-pin, to avoid the danger, when one charge was fired, of simultaneously exploding other charges. This groove is at present omitted, being judged unnecessary. The danger of such an explosion is now considered imaginary. You may discharge one chamber, and the contiguous ones, though you had filled them with loose powder, will not explode. The idea, also, that breech-loading is indispensable, if we would secure to a ball the maximum velocity and precision of aim, is no longer prevalent. The object of breech-loading—mainly to escape windage— is gained whenever bullets accurately fit the bore of the piece.

The principles which were at first developed in pistols were mostly applicable, and hence were soon applied by Col. Colt, to the fabrication of a dozen other species of arms; and, conversely, those which came to light in making muskets were usually suited to improve, in some particular, the model of the pistol.

That all the modern improvements which character-ized the arms manufactured by Col. Colt, originated with him, his friends will not maintain. But that he was the first man who made, and made known to the world, a six-shooter, rotating by the act of cocking, and which proved really serviceable, who will dispute? The multiform devices through which he strove to gain more and more perfectly this end, which before had been unattained by human wit, are conspicuous in the schedules accompanying his oft-reiterated applications for patents, and we give specimens of them in a note.

That he in fact invented what he claims as new, is evidenced in many cases by patents being granted him for the same, as well as by the injunctions and damages awarded him in courts of law when his patents were infringed. Thus, in 1854, it was testified concerning him, before a select committee of the British House of Commons, that he kept the trade in revolvers throughout America to himself, since by fighting them in law he had closed the Massachusetts Arms Company and every similar establishment.

It is told concerning a certain cider-seller at a country training, that, when some buyer grumbled at his price, since he knew that his was the only cask on the ground, he promised him all he could drink if he would show him another vender who sold cheaper. This the customer engaged to do, and so led the seller to the back side of the tent in which his barrel stood, and behold, there was a sharper, who, putting a gimlet through the canvas, and the barrel-head which rested against it, was slyly filling glass after glass, for which he asked only half price. All patentees in some way repeat that cider-seller's experience. The worst of it is, that they can seldom so soon detect the rogues who tap the other end of their barrels, or so easily thwart their tricks.

Imitation is frequently the sincerest flattery. Colt's contrivance of the lever ramrod was copied—as a man is mimicked by a monkey—by some Dutchman at Liege, so ignorant of its purpose that he made its arm too large for entering the socket in which it was intended to work. In attempting to load with this stolen lever we should be tantalized very much as if we were to rub wooden nutmegs on a grater, or to cut off rashers from a basswood ham.

In regard to the revolver as early as 1854, the London *Punch* pronounced it "the great American hit," and declared that *horse* pistols were everywhere being displaced by those of a *Colt;* while military magnates at Woolwich, the chief of British arsenals, bore the following testimony to Parliament, and that after applying as rigid tests as John Bull delights to use in proving —yea, "tenting to the quick"—whatever comes to him from Brother Jonathan, and claims to be something better than has been known in the old country. "It is our opinion," say they, "that Col. Colt's pistols are good, effective, substantial, and serviceable arms, and with moderate care and attention would answer all the exigencies of service." Four years later, when the weapon had been much improved, a pamphlet was printed containing one hundred testimonials, furnished by many more than that number of men, mostly military, who had themselves used the arm, and who agreed in holding it superior to all rivals of which they had any knowledge. Here was a chain of concordant testimony stretching through one-and-twenty years, and much of it given by such professional experts as the Texan or Western Rangers, Jack Hays, Joe Lane, Capt. Walker, and Ben. McCullough, or by officers who had examined, abroad as well as at home, the most efficient weapons the world could boast. Such authorities were not to be gainsaid.

Indeed, such a cloud of witnesses were already as needless as those collected in a similar book, not far from the same time, by the Bible Society,—holding a rushlight to show the sun,—that is, attesting the value of the Word of God. So at least this compilation seemed to those who remembered how often Colt's invention had been tried in actual warfare, and never found wanting—in Florida, Texas, and Mexico, not to speak of California, Australia, and the Crimea, and that twenty thousand of his weapons had been purchased by the United States Government, though it had two armories of its own.

Still, not a few testimonials accorded to Col. Colt—not evaporating in the mouth-honor of words, but crystallized in deeds and gifts—are so remarkable that they ought not to be all passed in silence.

One of those which he specially prized was a set of sleeve and vest buttons studded with diamonds, which an unknown Texan sent as a tribute of gratitude to him whose portable magazine, in the wilds of the far Southwest, had so often saved that borderer's life. Another was an unconscious but all the more expressive compliment. It lies in an anecdote often told in Washington with infinite relish by Henderson, when Senator there from Texas. Being solicited to plead the cause of a poor fellow who was charged with stealing a revolver, he was promised, in case of his success, the best fee he ever received. By special efforts he carried his point, and the accused was cleared. When his client, after acquittal, came to settle with his lawyer, he said: "Mr. Henderson, I have no money, but I am come to pay you what I promised—that is, the best fee you ever got; for here," said he, taking out a revolver, "is the identical pistol you have got me acquitted of stealing. I would not take any amount of money for it, but I have brought it to you as your fee."

Others of these grateful tributes, mostly from Japan, Siam, and outer barbarians, were specimens of national weapons, swords and spears, but especially the most uncouth and unwieldy of fire-arms. The heathen gods which we see set up in the missionary-house in Boston, flash upon us the contrast between the Gospel and Juggernaut; so these outlandish arms enable us to measure the perfection of Colt's arms somewhat as their donors measured it. A Caffre chief when he saw a plough, presented him by Queen Victoria, turning up its furrow, clapped his hands in royal rapture, exclaiming: "That plough is worth more than five wives." He who valued a plough so highly, what would have been his ecstasies over a revolver? To the bringer of such a treasure he would be likely to say, as Herod did to Herodias' daughter when her dancing pleased him: "Whatsoever thou shalt ask of me I will give it thee, unto the half of my kingdom."

Other presents, in return for specimens of his handiwork, came to Col. Colt from European potentates, kings that are kings, verifying the proverb of Solomon:

"Seest thou a man diligent in his business? he shall stand before kings." So costly, gorgeous, and exquisite are these oblations, that on a single one of them—a snuff-box from the Sultan of Turkey—the custom-house duty is reported to have been five hundred dollars; and that, as we survey them, especially the Russian gifts from Nicholas and Alexander, we fancy ourselves among the gold and diamonds in the green vaults at Dresden, or in that apartment of the patent-office where the presents to our government during fourscore years are reposited. Hither were they brought from as many and as widely severed nations as Portia's lovers, in the Merchant of Venice, pilgrimed through—

> "From the four corners of the earth they came,
> The Hyrcanian deserts, and the vasty wilds
> Of wide Arabia—as o'er a brook,
> From Tripolis, and Mexico, and England,
> From Lisbon, Barbary, and India,
> For the four winds blew in from every coast
> Renowned suitors."

We love to trace a stream to its fountain. Hence so many travellers, with endeavors no oftener baffled than renewed, have sought the source of the Nile. On a like principle, our minds run back from the fire-arms of our day to the earliest recorded or traditionary essays in that direction.

It was long customary for Germans to claim the invention of gunpowder. Their story is, that one Berthold Schwarz, of Mayence, where printing was invented, about the year 1330, chancing to mix nitre, coal, and sulphur, was pounding them in an iron mortar, when a blow of his pestle ignited the compound. According to another account, a spark falling into the mortar, the ingredients above named exploded, and threw the pestle, cover, and Berthold into the bargain, a long distance. Berthold, as a monk, was a mark for Protestant satire. Hence, one of their early poets thus sang concerning the origin of gunpowder and guns:—

> Dicite vos, Furiæ! qua gaudet origine monstrum?
> Nox genuit monachum, qui deinde Dæmone plenus,
> Protulit horrendum hoc primum cum pulvere monstrum.
> Machina, quam nullum satis execrabitur ævum, &c., &c.

The English assert that their Roger Bacon was acquainted with the composition of gunpowder a half century before Berthold's mishap. The ground of this assertion is the following words which occur in one of Bacon's works, *De Nullitate Magiæ:* "Take saltpetre, charcoal, and sulphur, and you can make thunder and lightning, *if you know how.*" The original phrase is, *si intelligis artificium.*

It is not impossible that Schwarz and Bacon were each independent discoverers of gunpowder, as Colt was of his revolver. Yet the researches of Conde among the manuscripts of the Escurial, near Madrid, have disclosed that that sort of powder was employed in fire-arms at Saragossa, in a battle between Spaniards and Moors, as early as the year 1118, that is, long before either Schwarz or Bacon was born; so that the opinion now prevails that fire-arms were introduced into Europe by Spanish Moors, and were borrowed from them by Christian nations. In all probability the Western Moslems derived them from their Eastern brethren, or from the Greeks of Constantinople. These Byzantine Greeks, in the judgment of the latest writers, had already for three centuries been familiar with gunpowder no less than with Greek fire. Thus, Prof. Draper, of New York, in his *Intellectual Development of Europe*, writes: "Of gunpowder, Marcus Græcus, whose date is probably to be referred to the close of the eighth century, gives the composition explicitly. He directs us to pulverize in a marble mortar one pound of sulphur, two of charcoal, and six of saltpetre. If some of this powder, says he, be tightly rammed in a long, narrow tube, closed at one end, and then set on fire, the tube will fly through the air. This is clearly the rocket. He says that thunder may be imitated by folding some of the powder in a cover and tying it up tightly. This is the cracker. It thus appears that fire-works preceded fire-arms."

There is said to be Chinese documentary evidence proving that explosive compounds were well known in the Celestial Empire before the Christian era. After all, however, the very earliest date to which we can trace the use of any explosive mixture analogous to gunpowder, we owe to one of the Greek classics. We there find Alexander the Great, about three hundred and thirty years before the birth of Christ, described as reluctant to attack the Oxydracæ, who dwelt near the Indus, because the gods had lent them thunder and lightning.

This loan of thunder and lightning John Bull has long desired to effect, and, according to a song in *Punch*, now sends for it this side the water, saying:

"Oh! Col. Colt, a thunderbolt
I'd buy for no small trifle;
But that can't be, and so let me
Get your revolving rifle."

to which the Yankees' response is—

"We'll come if paid, quick to your aid,
Six-shooters o'er the water," &c., &c.

Fire-arms, then, as we have them, are a legacy from
all the continents, and from more than one millennium
of time. Each nation has added something to their effectiveness. The Germans rifled the barrel, the Spaniards
added the bayonet, Italians devised the pistol, Dutchmen
the wheel-lock. A Scotchman, the Rev. Mr. Forsyth, obtained the first patent for percussion-locks no longer ago
than the year 1807. The profession of this patentee—
namely, the clerical—is worth notice, showing that here
as elsewhere, as, for instance, in originating the idea of
breaking the line in naval battles, the practical arts have
owed much to the meditations and researches of recluse
students, whose inventions have been at first received by the
self-styled practical men with ill-grounded and open con-
tempt. That the initial attempts to utilize the explosive
force of gunpowder would now seem ridiculous and
puerile is abundantly evident. Some of the first cannon
were only tubes of leather bound about with iron
hoops. In some antiquated but unaltered German
castles, specimens of this primeval sort may still be seen.
Others were such as we behold to this day on the Dar-
danelles, that is, horizontal holes bored in rocks, and
hence harmless save at the single moment an enemy
was passing their muzzles, unless, indeed, he were
obliging or chivalrous enough to halt in the line of
their fire.

Etymology, rightly viewed, gives out sparks of light
everywhere, so that in darkness we can always get
some guiding glimpses. The names of fire-arms are
eminently suggestive regarding their nature. A "fire-

arm" is one which produces fire. The largest pieces would seem to have been small, or they would not have been named *cannon*, a word which means a large reed, or *cane*. In the first guns the stocks formed one straight line with their barrels. The bending of the stock so as to enable the shooter to take aim is a German invention, and gave rise to the name *arquebuse*, a word which appears in Worcester spelled in eight different ways. This word, according to the Germans, signifies "bowed or hooked box." In their language a gun is still styled a "box," or a "fire-box," *Büchse*, *Feuerbüchse*. Our word blunderbuss is a corruption of *thunder-box*. The term *musket* originally meant a "hawk." That arm being invented in the palmy days of falconry, and bringing down its victims even from the air, as the hunting-hawk had done, was naturally called by the name of that bird whose exploits it achieved, and whose office in field-sports it soon took away, somewhat as among the Greeks the windlass, doing the work asses had done, was called a wooden ass, ὄνος ξύλινος. Through a similar association of ideas, the smallest piece of ancient ordnance was denominated a falconet, that is, a little falcon, and a larger species figures in old writers as a *falcon*. Worcester says that falcon in this sense is "a name formerly applied to a large cannon which carried a shot weighing seven hundred and fifty pounds!" In reading this definition we say to ourselves, O that we had such falcons nowadays! Soaring with supreme dominion they would, at one fell swoop, pierce the magazine of Sumter. Their calibre must have been well-nigh twice that of our fifteen - inch monsters, which weigh twenty-five tons.

The truth is, however, that mammoth guns were unknown until the nineteenth century. The marvel of
England ages ago,—familiarly termed by way of
meiosis,—"Queen Anne's pocket pistol," and said to
be inscribed,

> "From Calais to Dover is twenty miles over;
> Yet load me well and swab me clean,
> I'll carry a ball to Calais green;"

was after all only a ten-pounder. Yet another variety
of artillery was called in old English a "saker," which
is only a modification of the French name of a hawk,
sacre. Other sorts "of far-hissing tubes of death"—
as basilisks, culverins, &c.—derive their names from ser-
pents which were of similar shape, and fabled to thrust
out like tongues of flame.

Matchlocks were so called because touched off by a
match, that is, a bit of cord so prepared as to burn
slowly; wheel-locks took that name because fire was
struck out of their flints, or fire-stones, by means of a
tiny steel wheel whirling around against them.

It is often said that pistols owe their name to Pistoia,
an Italian city, where such hand-arms were early fabricated, as bayonets were named from Bayonne, in Spain,
because first made or used there. The most recent
etymologists, however, as Frisch and Diez, opine that the
word pistol is an off-shoot from the same root with
piston, and pestle, all at first meaning a pounder or
bludgeon. A new implement not unlike a bludgeon in
shape and size, and which did its work, might well
take its name. British writers in Notes and Queries,
have cited the word pistol as used by Strype as
early as the year 1575, but judge that in 1541

that word had not been imported into England, because it does not occur where, if current, it would naturally be used, that is, in the statute of that year which prohibits the possession of any "hand-gun, hagbut, or demihake, other than such as shall be in the stock and gun [that is, in entire length] of the length of three-quarters of one yard." In the passage from Strype, the word "dag," the early form of our dagger, being used as synonymous with pistol, somewhat confirms the theory that pistol at first meant a bludgeon.

But the name *gun*, which is abbreviated from "engine," is the most significant term for all the larger fire-arms except mortars. It expresses men's feeling that a contrivance which will throw balls, and that by means of explosive compounds, is, by way of eminence, the engine, the chiefest among the destructive fruits of human ingenuity. Hence the element, gun, enters into many derivatives and compounds, such as "gun-metal," "gun-shot," "gun-cotton," "gun-powder," "gun-smith," "gunnery," &c. The name "artillery," which is related to the word *art*, owes its origin to the great degree of art, or skill, needed for making and managing that arm of military service.

Such are some names of fire-arms; such is in outline their history; such are some of their characteristics. We often hear it asked: Are they a blessing, or a curse? According to our answer to this question must we view him who improves them as a benefactor, or a malefactor, of his race. Regarding the value of this invention, the strains which poets have sung are rather depreciatory. Milton traces the first idea of cannon to Satan himself.

> " For whence
> But from the author of all ill could spring
> Malice so deep?"

Nor yet does he deny but that some man, ages afterward, might become an independent inventor of what the arch-apostate had so long before excogitated, for he says:

> " In future days, if malice should abound,
> Some man intent on mischief may devise
> Like instrument to plague the sons of men."

In a similar vein, Shakspeare introduces a carpet-knight, insisting:

> " That it was great pity—so it was—
> That villanous saltpetre should be digged
> Out of the bowels of the harmless earth,
> Which many a good tall fellow had destroyed
> So cowardly; and but for these vile guns
> He would himself have been a soldier."

Cervantes shows us Don Quixote exclaiming: " Happy were the ages past while strangers to these infernal instruments of artillery, the author of which is, I firmly believe, now in hell, enjoying the rewards of his diabolical invention, that puts it in the power of an infamous coward to deprive the most valiant cavalier of life; for often in the midst of that courage which fires the gallant breast, there comes a random ball, how or from whence no man can tell, shot off, perhaps, by one who fled and was afraid at the flash of his own machine, and puts an end to the existence of a man who deserved to live for ages," &c. In a similar spirit Rabelais represents Gargantua, remarking that " printing was invented in his time, as a counterpoise to the diabolical suggestion of artillery."

The truth is, that a like prejudice has at every step beset improvements in every sort of arms. Thus, one of the most ancient of Greek anecdotes tells us of a philosopher who, gazing at the gigantic arrow, huger than a weaver's beam, that was to be projected by a catapult, which is, literally, a "knock-down," and which may be described as a bow longer than the yard-arms of a ship, and bent by a sort of windlass, cried out: "Behold the grave of valor!" His idea was, that against such a foe valor would be useless, and that in using that weapon it would be needless. In the course of time, however, it appeared that the sage was every way mistaken. When catapult was pitted against catapult, the users of each were in as great danger as the archers of old had been; and again brave men arose who, counting darts as stubble, stormed catapult batteries. When wit and science have done their utmost, it still remains true, that—

"A man's a man for a' that."

Moreover, it has long been agreed by historians, that gunpowder has diminished the loss of life in war. One reason is, that fire-arms decide the fate of battles while the combatants are still so far apart that the vanquished have time for flight, whereas, in ancient battles, the defeated army was usually annihilated. The perfection to which guns have recently attained makes their inevitable overthrow, if they stand, manifest to the weaker party, while they are further removed from their victors than in conflicts a generation ago. The consequence is, that those worsted now escape, or surrender, in cases when, on the ancient system of warfare, they

would have perished on the spot. The most perfect
of weapons, then, best deserves the name of Peace-
maker. Once render your arms so perfect that you will
be sure to annihilate your antagonist, and you will have
none. Let him and you be both so well armed that
you will destroy each other as utterly as those Kilkenny
cats, which ate each other up all but the tails, and you
and he will both be slow to engage in a duel. Those
who live in glass houses beware of throwing stones.
Through convictions that these things are so, the Italians
have a proverb: "One sword keeps another in its scab-
bard;" and the French say: "He who bears the sword
bears peace. *Qui porte épée porte paix.*"

Again, the repeating pistol is a weapon too elaborate
to be manufactured except in civilized states. Therefore it must be slowly introduced, and always scarce
among barbarians, and even among semi-civilized races.
Hence, it cut off half from the length of our war on
the Seminoles, in Florida. Arms on that principle
hastened the independence of Texas, perhaps achieved
it. They doubled our triumphs in Mexico, while they
halved our losses. They have given savages, the world
over, a dread of harming the miner and the pioneer,
never felt before. They have struck terror into the
Bedouin banditti, who, from time immemorial, had plun-
dered at will Christian travellers in the Orient. If Captain Cook had carried a revolver on his voyage of
discovery, he would not have perished as he did. No
savage offered him violence till he had fired the last
charge from his double-barrelled gun. The more, then,
arms are perfected, the stronger is civilization against
outer barbarians.

"No more on polished nations from afar
Shall Scythia pour the purple cloud of war,
Till, torn and mangled 'neath her savage sway,
Their laws, their arts, their temples pass away."

We may hold the repeating fire-arm a blessing to mankind, since it gives protection where it is most needed—that is, to the weak against the strong. No dwarf but can wield a revolver, no giant but that it can kill. Indeed, this invention gives the dwarf some superiorities over the giant, no matter what his thews and sinews. The bigger the giant's bulk, the larger mark is he, and hence the easier to hit. The six-shooter also insures to mind, "a little body with a mighty heart," its rightful supremacy over matter. If David has more mind—of which courage is one form—than Goliah, he will be sure to slay him, for he will not only have a larger mark, but he will take a more unerring aim.

Yet once more: a principal use of pistols has always been to help officers—whether naval, or military, or civil —enforce obedience upon those rightfully under their authority. The elder Adams complains in his diary that the ship of war which took him to France would be good for nothing in action, because the officers had few pistols for forcing the men to stand to their guns. Now for this purpose, whether on sea or land, as well as to quell mutinies and uphold discipline, the arms of to-day are beyond description superior to those of our fathers. In general, then, police must become more efficient; rightful authority more triumphant over anarchy; armies better trained, more sure to execute orders, and hence more invincible.

It is obvious that improved arms have been the instruments of vast evils. So has fire. How oft the sight
24

of means to do ill deeds makes ill deeds done. Whatever may be greatly used, may be greatly abused—and is so. Few proverbs hide a truth deeper and wider than the scholastic maxim, *corruptio optimi pessima.* The worst things are the best perverted. As a good woman is diviner than a good man, so an evil woman is more fiendish. It was no common angel out of whom Satan was made. Throughout the universe, he that will may gather illustrations, that the better a thing is, the worse are its corruptions. What better than religion? What worse than hypocrisy?

But if whatever has been greatly abused may be greatly used, we ought to use it, and unfold its capacities for good. So thought Wesley, when he said Satan had had the good tunes long enough, and so reclaimed them from bacchanals for his village hymns.

But whether we view improved fire-arms as a blessing or a curse, all must admit that, they being once invented, all good men, and governments, so far as emergencies demand, should avail themselves of them. Logical fallacies are evils, but when once machinated and published, they are sure to be employed by those skilled to make the worse appear the better reason; and therefore, to put us on our guard, they are inserted in all logics. In like manner, improved arms will be sought for by robbers, assassins, and all disturbers of public peace. They must therefore be turned against such evil-doers, by all who would uphold order, political and social—the first law of Heaven, without which force is the only right. They will be abused for attack, therefore they must be used for defence. Through our ignoring, as a nation, this plain truth, we fell down so low at the outset of the Southern rebellion, bloody treason flourished over us, while our Government was glad to buy, in European markets, genuine quaker guns, rifled with a file two or three inches from their muzzles, and in which the cone nipples were not pierced with any holes at all. In truth, if the sentimentalism of pseudo-philanthropists had had a little more free course, ours would have been the condition of the land of Israel when "there was no smith found throughout it, but all the Israelites went down to the Philistines, to sharpen every man his share, and his coulter, and his axe, and his mattock; so it came to pass in the day of battle, that there was neither sword nor spear found in the hand of any of the people." But every civilized state—and that the more, the more each grows in civilization—acts on the maxim that no duty is more imperative, and no policy more wise, than to pre- pare for war in time of peace.

> " Hence warlike arms in magazines we place,
> All ranged in order and disposed with grace;
> Not thus alone the curious eye to please,
> But to be found, when need requires, with ease."

If, then, as things are, the peace of the world demands the multiplication of fire-arms, and those of the best models, we cannot fail to view with peculiar interest the manufactory which stands first and foremost among all those ever established, either in this country or in any other, by private enterprise, for the fabrication of such weapons—nothing less than a palace of industry.

The following description was written in the autumn of 1863. Had it been deferred a few months later, it could not have been made at all, since, within the next half year, those portions of the establishment which are most particularly treated, were laid in ashes. To not a few will it be a lasting satisfaction that the shadow of this catastrophe, preceding the event, or some other pre- sentiment, hurried on the completion of this memorial of a structure which, while appearing built for monumental endurance, was in truth a vapor ready to vanish away. Ephemeral as the present publication may prove, it already, in contrast with the fire-scathed ruins, confirms the saying of Hazlitt :—"After all, words are the only things that endure here on the earth."

Before proceeding to the description of the Armory, it will be of historical interest to mark, by illustrations drawn from the specimens patented, the gradual develop- ment of the Colt Revolver.

DEVELOPMENT OF THE COLT REVOLVER.

THE PARTS AND COMBINATIONS.

A revolving pistol consists essentially of, 1st, The Barrel; 2d, The Cylinder; 3d, The Lock-frame (containing the lock); 4th, The Stock.

1. The barrel is made of steel and is rifled. It is in all respects like the barrel of any muzzle-loading arm, except that it is open at both ends, and on its lowest side are a socket for the rammer, and the fixtures for fastening the rammer and its lever to the barrel. These fixtures are all forged in one piece with the barrel. There is also a slot below for holding the key of the cylinder-pin, hereafter described.

2. The cylinder is a piece of steel in which five or six chambers parallel to the axis are bored. Their bore and that of the barrel are the same. They are open to the front, and stop at a distance from the rear of the cylinder great enough to leave sufficient metal behind the chamber to give proper security against bursting. Behind each chamber, and entering it, an orifice is cut, which the screw on the base of the cap-cone fits, so that the cone is fixed directly in rear of the chamber.

Besides the chambers, there is another hole in the cylinder whose axis is the axis of the cylinder, and which is bored entirely through it. Through this, and fitting it precisely, passes a pin from the lock-frame to the barrel. The pin is parallel to the bore of the barrel, and so far below it that the revolution of the cylinder brings in succession each chamber directly behind the

barrel, so that the chamber and bore of the barrel can
be made continuous.   This pin secures the cylinder in
position between the lock-frame and barrel, allowing it
only to revolve about its axis.   It is secured to the
barrel by a key passing through a slot cut in the pin,
and a corresponding one in the barrel.

On the rear of the cylinder is cut a ratchet, having
five or six teeth, as the cylinder has five or six chambers.   The centre of the ratchet is on the axis of the
cylinder, and the teeth are so arranged, that when the
piece is at full cock a chamber is directly in rear of the
barrel.

On the surface of the cylinder are cut as many small
slots as there are chambers.   The lowest of these slots
is entered by the end of a bolt, which is movable by
the action of the lock, and is pressed into the slot by
a spring constantly acting.   So long as the bolt is in
the slot, the cylinder is immovable.

3. The lock-frame is directly in rear of the cylinder,
and consists of the recoil-piece, into which the cylinder-
pin is fastened; the lock, which contains the machinery
for exploding the cap, as well as revolving and locking
the cylinder; and a frame, which contains and holds in
place the mainspring.

In the lock the sear and trigger are in one piece, as
are also the hammer and tumbler.   In these respects the
lock differs from that used before the invention of the
revolver by Colt.   The mainspring acts upon the tumbler and hammer directly.

The tumbler has fastened on its face a " hand," which
engages the ratchet on the rear of the cylinder, and is
held against it by a spring.   It also has a projecting

pin, which is so arranged that at the proper time it engages the bolt which locks the cylinder, lifting it out of its seat in the cylinder-slot, and giving it freedom to move under the action of the "hand." When the pin no longer acts upon the bolt, the spring, which constantly acts upon it, forces it into the first slot which it meets in the revolution of the cylinder, thus locking it.

The action of the lock is as follows. The hammer is supposed to be resting upon one of the cap-cones.

The hammer being slightly raised, there is at first no motion of any other part, except that of compression of the mainspring. This device is used to permit any pieces of exploded caps to fall out at the raising of the hammer, before the other parts are in motion. The raising of the hammer being continued, the pin in the face of the tumbler disengages the bolt from the slot in the cylinder; immediately afterwards the "hand" engages a tooth of the ratchet, and as the hammer is raised the cylinder is revolved by the "hand" one-fifth or one-sixth of a revolution, according as there are five or six chambers. When the hammer arrives at full cock, the tumbler-pin is disengaged from the bolt, which flies back into the cylinder-slot by the action of its spring, and bolts the cylinder. The piece is then ready for discharge, and the cap can be exploded by pulling the trigger.

4. The stock, composed of wood or ivory, is immediately in rear of the lock, and embraces that part of the lock-frame which contains the mainspring. It forms the handle or grip.

The piece is loaded by inserting the powder and ball or cartridge in the front ends of the chambers. These are successively brought under the rammer, and the loads are pressed home by the action of an ingenious lever arrangement attached to the barrel and worked by the hand. The piece is then capped, when it is ready for discharge.

Although, in the various forms and models of revolving arms made since the date of Colt's inventions, there have been some deviations from the details of this description, the principle of the arm is in all of them the same as that given here.

VARIETIES.

The TEXAS PISTOL, as it was at first named, was made at Paterson, New Jersey, about 1838. It was the first form of revolver which came at all into general use, and was a popular arm along the Western frontier. Its calibre was .34-inch. It had a concealed trigger, which was thrown out by the act of cocking, was without a guard, and had no lever attachment for loading.

During the Mexican war, in 1846–'47 and '48, this pistol was in great demand in Texas and Mexico, and one hundred dollars in specie was not an uncommon price for it, even after it had been long in use.

Another pistol similar to the TEXAS ARM was also made at Paterson. It was called the Walker pistol by Col. Colt, out of compliment to a distinguished Texan ranger of that name, with whom the pistol was a great favorite.

It was much larger and heavier than the Texas arm, and, although it differs in proportion of parts, was in form and arrangement very similar to the pistol known as Colt's Old Model Army Pistol.

The lever and rammer were attached to this pistol,

WALKER PISTOL.

TEXAS ARM.

OLD MODEL NAVY PISTOL, WITH CARBINE ATTACHMENT.

TEXAS ARM.

though the idea of a lever rammer was suggested by sketches of a much earlier date. Its calibre was .44-inch.

The OLD MODEL ARMY PISTOL followed the Walker pistol in 1847, and was extensively introduced as a weapon for cavalry. The calibre is .44-inch, and weight 4 lbs.

POCKET REVOLVERS, with 6-inch, 5-inch, 4-inch, and 3-inch barrels, and .31-inch bore, were introduced about 1848. They were at first made without lever rammers, and were loaded by removing the cylinder from its pin, and using the pin as a rammer.

THE OLD MODEL NAVY PISTOL, which has been the most popular of all the pistols, was introduced about 1851. It was a great advance upon its predecessors, containing an improvement in the form of cylinder-slot, which, with the bolt, secured the cylinder during discharge, making the action of the bolt more certain. This improvement was patented in 1850. The same patent also covers the safety-pins inserted in rear of the cylinder, which added much to the efficiency of the arm. Both of these improvements have been retained in nearly all of the models made since.

Its calibre is .36-inch, and weight 2 lbs. 10 oz.

The OLD MODEL POCKET PISTOLS were improved in 1849 by the introduction of the perfected features of the Navy Pistols, and have not been materially altered since.

Their calibre is .31-inch, and weight from 24 to 27 oz., dependent upon the length of the barrel.

A small NEW MODEL POCKET PISTOL was introduced in 1855, and has an entirely different arrangement from those which preceded it. The cylinder is inclosed by a

25

metallic frame, the trigger has a different guard, and the action of the "hand" is different.

The calibre is .31-inch, and weight 1 lb.

The Rifle made at this time is similar in its arrangements to this pistol.

The NEW POCKET, and NEW MODEL POLICE, NEW MODEL ARMY, and NEW MODEL NAVY, have all been introduced since 1860. They present nothing new in principle, but by a better arrangement of parts, and in the Army Models by less weight, suit the markets better. The calibre, too, of the Pocket and Police pistols is .36-inch, the same as that of the Navy pistol. The increased calibre makes them more deadly weapons. The calibre of the New Model Army Pistol is .44-inch.

The New Pocket Pistol weighs from 25 to 28 oz.

The New Model Police Pistol weighs from $24\frac{1}{2}$ to 26 oz.

The New Model Navy Pistol weighs 2 lb. 10 oz.

The New Model Army Pistol weighs 2 lb. 11 oz.

There are now made at the works fourteen models of pistols, differing essentially from each other, varying in calibre from .31-inch to .44-inch, and in weight from 1 lb. to 2 lb. 11 oz.

The OLD MODEL RIFLES were made in Paterson in 1836–'37. They were at first made to cock by a lever under the lock-frame. They were loaded by a detached lever and rammer, the end of the lever fitting into a slit in the frame, which formed a fulcrum, and the balls were pushed home by the action of the lever on the rammer. Afterwards the lever was attached to the side of the barrel, and the hammer was arranged as it is in the pistols.

Rifles for sporting purposes are now made of calibres .36, .44, and .56-inch. In their arrangement they are like the New Model Pocket Pistol previously described. They were introduced in 1856–'57 and 1858.

The barrels are of various lengths, from 24 to 37½ inches.

SHOT GUNS of .60-inch and .75-inch calibre, and of lengths from 27 to 36 inches, are made now, and are averaged like the Sporting Rifles.

DEVELOPMENT OF THE REVOLVER.

In the year 1830, on a voyage to Calcutta, as a sailor before the mast, Samuel Colt whittled out of a ribbon-block the first model of the cylinder of his revolver, and reduced to a tangible shape parts of the machinery by which the weapon was made effective.

In the Summer and Fall of 1831 he constructed at Hartford his first two pistols, one of which burst upon the first trial. In 1832 he deposited a description of his invention in the Patent Office at Washington; and in 1833, constructed in Baltimore both a pistol and rifle on the principle, which was patented in England and France in 1835, and in the United States, Feb. 25th, 1836. With the coöperation of some New York capitalists, the "Patent Arms Manufacturing Company" of Paterson, N. J., was chartered by the New Jersey Legislature on March 5th, 1836, its first directors being Thomas A. Emmet, Daniel K. Allen, Elias B. D. Ogden, Daniel Holsman, and Elias Vanarsdale, Jr. Its proposed capital was two hundred and thirty thousand dollars, and subscription books were opened in April, 1836;—it is probable that only one hundred and fifty thousand dollars were ever paid in. He assigned his letters patent to the company, and undertook to give his time and attention to its business, receiving therefor a fixed price, varying from one to two dollars, for each arm manufactured. His proposition of March 9th, 1836, was to assign the patent, and the right to all improvements in arms and machinery; to devote not over nine months to setting the factory in operation, and to give such future aid as might be needed in perfecting the improvements, at a salary of one thousand dollars per annum—the company to establish the works within six months, and extend them as might be needed; not to dispose of any right without his consent, to pay him semi-annually half the net profits of the company, and half the proceeds of sale of any right, six thousand dollars being advanced in six monthly payments, and the right of subscription to fifty thousand dollars of the stock being reserved to him for a year. His next step was to prevail upon the United States Government to adopt the arm; and in accordance with a resolution of the Senate of January 21st, 1837, a trial was made of the weapons of Hall, Cochran, Colt, and Baron Hackett, by a board of officers appointed by the Secretary of War, viz., Brevet Brigadier-Generals J. R. Fenwick and N. Towson, Colonel G. Croghan, Brevet Lieutenant-Colonel Worth, Lieutenant-Colonel Wainwright, Brevet Major Baker, Captain B. Huger, and First Lieutenant R. Anderson. They reported, September 19th, 1837, it as their unanimous opinion, "that from its complicated character, its liability to accident, and other reasons, this arm was entirely unsuited to the general purposes of the service." In October, 1837, he received a gold medal from the American Institute, at its fair in

New York, and was elected a member. But the opposition to the weapon on the part of both army and navy
officers was so great that the Government could not be
induced to use them, and the cost of manufacture was
so high that but small private sales could be effected.
The arms that were manufactured were, however, all
disposed of, many of them at very reduced prices, to
the Texan and frontier pioneers, and in the hands of
the Texan "rangers," under Hays and Walker, had much
to do in effecting Texan independence. Soon after the
breaking out of the Seminole war, in 1838, he went to
the seat of the war, exhibiting his invention at Charleston, S. C., on his route (February 19, 1838); and on
the 9th of March, at the request of Lieutenant-Colonel
W. E. Harney, a board of officers, consisting of Captains
W. W. Tompkins, W. W. Fulton, and J. Graham, was
appointed by Colonel D. E. Twiggs at Fort Jupiter,
East Florida, to examine and report upon the efficiency
of his rifles. They reported on the same day very favor-
ably, and recommended the partial adoption of them in
one of the dragoon regiments. Harney forwarded the
report, with the annexed recommendation of Colonel
Twiggs and himself, to Major-General Jessup, who ordered the purchase of fifty, of which General Harney,
in November, 1850, says:—"They were placed in my
hands. They were very delicately made, easily put out
of order, and very difficult to repair. They were the
first ever used or manufactured. Thirty odd of them
were lost at Caloosahatchie, all at the time in good order, and most of the others, with little repairs, would be
still good, and were turned over to the ordnance officer
at Baton Rouge in 1842 or 1843. During the whole of

this time they were in constant use, and not one single
accident to the injury of any person occurred; and I
honestly believe that but for these arms the Indians
would now be luxuriating in the everglades of Florida."
On the 26th of October, 1838, another trial of his
arms was made in New York, before the American Institute. In 1839 a second patent was taken out, covering several improvements, the chief of which was the
loading lever. In pursuance of a Senate resolution of
March 17, 1840, a board of navy officers (Captains C.
S. McCauley, J. H. Aulick, and L. Twiggs) was appointed to report upon the merits of the pistols and
rifles as an arm for the use of boarders and marines.
On the 6th of May they reported, after trials made at
Washington, that "their advantages were counterbalanced
by complexity of · construction, and consequent greater
liability to derangement and accidents," but recommended
them for arming boat expeditions, and acknowledged the
great superiority of the percussion to the flint lock. In
accordance with another resolution of July 20, 1840, a
second board of dragoon officers (Captain E. V. Sumner,
First Lieutenant W. Eustis, and First Lieutenant H. S.
Turner) at the Dragoon School of Practice, Carlisle Bar-
racks, Pa., examined into their fitness for the dragoon
service, and also the merits of the water-proof ammunition,
invented by Mr. Colt. The board recommended a trial
of the repeaters by the arming of a company or two
for six months in field and garrison, and also advised the
introduction of the water-proof ammunition. In consequence of this report, the Government purchased one
hundred and sixty carbines in March and July, 1841,
at forty-five dollars each.

The following table exhibits the number of Pistols, Rifles, and Muskets manufactured by Colt's Patent Fire-arms Manufacturing Company, from January 1, 1856, to December 30, 1865:—

YEAR.	PISTOLS.	RIFLES.	MUSKETS.
1856	24,053		
1857	39,164		
1858	39,059		
1859	37,616		
1860	27,374		
1861	69,655	3,193	
1862	111,676	2,287	8,500
1863	136,579	1,213	49,844
1864	10,406		46,201
1865	58,701		9,435

The following table of Government orders for Colt's Repeating Fire-arms is the best evidence how steadily they grew in estimation in proportion as they were tried. These orders amounted to about four hundred thousand dollars within ten years, and it is noteworthy that those for nearly half that sum were issued during the last year which the table covers. It need scarcely be added, that during the last decade the Government orders have largely exceeded those in all former years. In proof that Colt's arms were still more highly valued abroad than at home, it is enough to mention the fact that, as early as 1854, he had sold the Viceroy of Egypt five thousand revolvers, and the incredible bill of forty times as many to the British Government, then in the midst of Crimean campaigns.

UNITED STATES ORDERS—WAR DEPARTMENT.

1841—March 2..........	100	Carbines...............	$45	....	$4,500.00
July 23	60	...."..................	"	....	2,700.00
1847—January 4.......	1,000	Holster Pistols........	28	.	28,000.00
November 2 1,000		"	"	....	28,000.00
1849—January 8..	1,000	"	"	 25	25,000.00
1850—February 4	1,000	"	"	.... "	25,000.00
1851—May 8..... . ..2,000		"	"	 24	48,000.00
1853 " 26..........	1,000	"	"	.... "	24,000.00
1855—January 15	1,000	"	"	.... "	24,000.00
July 27..........	1,000	Belt	"	.... "	24,000.00
December 6......	100	"	"	.... "	2,400.00
1856—February 13	200	"	"	.... "	4,800.00
April 21	370	"	"	.... "	8,880.00
" 23	125	Holster	"	.... "	3,000.00
" 26........	100	"	"	.... "	2,400.00
" 26........	100	Belt	"	.... "	2,400.00
" 29........	55	Holster	"	.... "	1,320.00
May 3........	50	"	"	.... "	1,200.00
June 10	125	Belt	"	.... "	3,000.00
August 14......	6	"	"	.... "	144.00
September 19 ...	60	"	"	.... "	1,440.00
1857—January 7	101	Rifles................	50	....	5,050.00
" 7	500	Belt	"	 24	12,000.00
April 13........	1	"	"	.... "	24.00
May 13.........	250	"	"	.... 20.49......	5,122.50
" 18..	16	"	"	.... "	327.84
June 2	100	"	"	.... "	2,049.00
" 4........	10	"	"	.... "	204.90
" 18	200	"	"	.... "	4,098.00
July 11.........	150	"	"	.... "	3,073.50
August 12......	100	"	"	.... "	2,049.00
" 14......	300	"	"	.... "	6,147.00
" 20......	200	"	"	.... "	4,098.00
September 5.....	150	"	"	.... "	3,073.50
November 3.....5,000		"	"	.... 18.47½.....	92,395.83
" 21.....	300	Rifles................	42.50...		12,750.00
	17,829				$416,567.07

NAVY DEPARTMENT.

1852—July 10..........	25	Army Pistols........	$25...........		$625.00
" "	50	Navy	"	.... "	1,250.00
" "	13	6-inch	"	.... 19.30........	250.90
" "	6	5-inch	"	.... 18.30........	109.80
" "	6	4-inch	"	.... 17.30........	103.80
1856—June 16........	50	Navy	"	.... 18.64½........	932.25
1857—May 21...	50	"	"	.... 19.43½.....	971.75
June 16........	30	"	"	.... "	583.05
August 3........	50	"	"	.... "	971.75
September 28....1,870		"	"	.... "	36,843.45
	19,979				$458,708.82

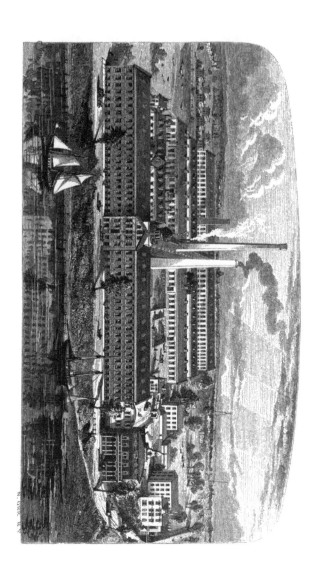

THE ARMORY

THE ARMORY.

THE ARMS FACTORY has an aim in the same general direction with the greatest industrial movements of the nineteenth century. All these are for the million. Invented and organized by a few, they are carried on by the many, and inure to their benefit. Thus, common schools, good enough for the best, yet cheap enough for the poorest, are for their education; the power-loom and sewing-machine, lavishing garments which the Chinese call "heaven-woven," and "heaven-made," are for their clothing; the printing-press is for their reading, which is travels at home; steam is for their travels abroad; and the arms factory is that they may have guns "which are guns," such arms as are a people's best peacemakers— being death to him who breaks it.

It is eminently an iron-work. Its machines, tools, stock, products, are all mainly of iron, which is the metal of civilization. The gorgeous East showers on her kings *barbaric*, pearl and gold, but not much of iron. No fragment of that truly precious metal was ever smelted in any savage tribe; the use of it begins with the first step of national culture, and from first to last goes on increasing with every subsequent step of progress.

However poets, who delight in fiction, sing of a golden age as the ideal of excellence, philosophers, being lovers of truth, must hold that no name but "iron age" befits that glorious era. Homer, though blind, may have had some insight into this fitness of things, for he gave the goddess of love and beauty to the blacksmith god, while he doomed Jove, who despised the hammer, to the caudle lectures of a shrew.

Col. Colt's years, after obtaining his first patent, in 1836, were twenty-six, the first twelve of failure, the last fourteen of success. We ought not perhaps to say "failure," although they involved the loss of a quarter of a million dollars by the Paterson company before 1842, his own pecuniary ruin, and inability for six years thereafter to find capitalists who would risk a dollar in carrying out projects, which seemed proved by costly experiments as impracticable as twisting ropes of sand, or weaving cobwebs. These years were not failures, for they demonstrated, to those best able to judge, the superiority of his weapon, while still as it were in swaddling-clothes, and they doubled its perfection. While the world gave him, as a ruined man, the cold shoulder, he scarcely noticed the neglect, for he knew that diamonds are always found in dark places, and was applying his principle a thousand ways, that he might make his own the very best of them all. That ideal was in him, and it burned him like fire, till he had brought it out, realized it, and "a fabric huge rose like an exhalation," to scatter it abroad among men to the ends of the earth.

His experience, then, had much in common with that of other millionaires. Such men make a little slowly—but much fast. They so make a little that they learn to

make much.   He was not unlike the merchant who sold
all that he had to buy the most precious of pearls, which
he then brought to Madrid for sale.   When asked by the
courtiers how he dared to invest his whole estate in one
single jewel, his answer was: " Because I knew there was
only one such pearl, and also that there was a king of
Spain in the world."   His words were so acceptable to
Philip II., that he paid the adventurer a fabulous price
for the unique gem, which, styled La Peregrina, or the
Fair Pilgrim, is still admired in the Spanish capital.

Col. Colt's last fourteen years were such a flood-tide of
prosperity, setting in after a lowest ebb, as has seldom
rolled its waves anywhere.

> " Like to the Pontic sea,
> Whose icy current and compulsive course
> Ne'er knows retiring."

His order from the United States for one thousand
pistols, which he made in 1847, at Whitneyville, Connecticut, was vouchsafed him because the merit of his
work had been tested, and was undeniable.   As a natural consequence, other orders followed, and capital was
afforded him to commence a factory in Hartford.   His quiet
assurance, that wherever his arms should go they would
prove their own best advertisement, oozes out in the
words he stamped on every barrel—" Address Col. Samuel
Colt, New York, U. S. America," as if no one could
see, without seeking for, such an arm.   That assurance
was not presumptuous.   No year since has passed without
a war somewhere in the world, calling for improved arms.
Every peace also, scattering disbanded soldiers through
communities, has wakened the fears of the timid, and made
them eager purchasers of the best life and property

preservers. Again, a recent letter from Iowa says: "The wives of copperheads go barefoot, that their lords may own revolvers—and Union men keep up with them." Further, in regard to California and Australia, it may be doubted whether either of them enriched any pilgrim to its auriferous placers, more than it did the improver of the pistol, through opening unexpected markets for his wares.

The site of the armory founded by Col. Colt is admirable. It is within a short walk of the State House, railroads, and the business centre of Hartford. It has at hand all the water required for its manifold necessities; it is close to a navigable stream, so that coal, iron, and all stock, can be landed at its doors, and its products, whenever it may be desirable, can be shipped by the Connecticut. But this ground, within the corporate limits of the city, and amounting to two hundred acres, could have been purchased a dozen years ago for no more than some single acres within a mile of it could then have been sold. How was this possible? Why, the tract where arms are made, and where a thousand people now dwell, had lain for centuries without inhabitant or house. The reason was that the river, swollen by spring rains and the melting of northern snows, overflowed all this region annually with a Nile-like inundation. One is tempted to think that the Dutch, when all the land was open to them, in a fit of home-sickness chose as their place of settlement the nearest eminence to this flooded field, because its half aquatic surroundings reminded them of their father-land, that was so often submerged.

In this amphibious flat—where land and water held such divided empire—at best but mediocre grazing ground, or where crops were in danger of being harvested by

autumnal floods—to discover the situation most convenient, economical, and every way best adapted for an
arms factory, was a mark of no small acumen. The truth
is, it was not by any means *found* such a situation. It
was *made* so by herculean labors, in spite of obstacles
which many expected, and some, it is to be feared, like
other prophets, wished, would prove insurmountable. Possibly it was thoughts of the old Dutch squatters which
suggested to Col. Colt the practicability of reclaiming
what through so much of the year was no man's land.
However this may be, he in fact built a dyke, in genuine Dutch fashion, one which rouses in every transatlan-
tic tourist reminiscences of the rampart against the wild
and wasteful German Ocean, which moved his wonder
wherever he landed, between France and Denmark. This
embankment, reared at an expense of one hundred and
twenty-five thousand dollars, is on an average one hundred feet thick at the base, forty at top, and two miles
long. The original purpose was to grade it thirty feet
above low-water mark. But when the Connecticut, during the progress of the work, rose higher than ever before, as if trying it with a proof charge—or indignant
at being thus "put into circumscription and confine," the
grade was made a little higher, and fixed at thirty-two
feet. In the height of this unprecedented freshet, many
a croaker said to this mound-builder: "Your mud heap
will wash away now."—"Very well," was his answer, "I
am off for New York to-morrow, and will catch it as it
goes by, and bring it home again in my trunk." In
point of fact, that mound then stood like a pyramid. It
was soon set thick with willows, the roots of which form
a wattled wall, year by year waxing stronger and stronger.

These trees, while buttressing the water-wall, were not only an ornament and a shade, but yielded material for a new manufacture, namely, willow-work, which forty imported artisans soon bent and twisted into as many curious shapes as we see in the kaleidoscope, and those all no more curious than useful. Their planter thus contrived that they should pay a double debt, like that old wall in Aphek, which, falling on the soldiers of King Benhadad, gave them, as Fuller phrases it, not only death, but gravestones at the same time. The earth for forming the dyke was excavated and drawn up from the shore and bed of the river itself. The stream, having thus a deeper channel, was less likely to overflow its banks, and was most adroitly made to rear a barrier against its own assaults, somewhat as a victim of crucifixion is forced to become his own tormentor.

Hollanders are wont to say that their highest national eulogy is compacted in this single hexameter line:

"Tellurem fecere Dii, sua litora Belgæ."

"The gods made the world, but the Dutch made their own land." A similar boast beseems him among us who not only devised a new arm, and built a factory with machines for fabricating it, but also created terra firma as a stand-point for that factory. Archimedes begged for a stand-point; Colt made one.

While thus helping himself by dyke-digging, Col. Colt helped so many of his neighbors, that, seeing their lands no longer invaded by water-raids, they would surely concur in the rather Hibernian compliment which the Scotch pay to that British officer who first opened a road over their highlands, namely:

"If you had seen these roads before they were made,
    You would lift up both hands and bless General Wade."

Mrs. Sigourney, whose writings and whose life are equally full of genial appreciation for great deeds, no matter in what sphere of enterprise, has celebrated the dyke as freshet-proof, and the retreat of the disheartened River, when,

"At length, all baffled, he drew away
His marshalled hosts from the bootless fray,
Retreating sullenly and slow,
With muttered words of shame and woe;
So Xerxes turned with fallen crest
From conquering Greece of the dauntless breast;—
For one who erst in his boyhood's hour,
    Sported amid yon hillock's sheen,
Had vanquished the Flood whose beauty and power
    Were the pride of his native valleys green."

The arms factory is a pile of building surpassed in magnitude by few industrial establishments. The ground plot of the original edifice had the shape of the capital letter H. One of the upright strokes, which were each five hundred feet in length by sixty in breadth, faces the willow-fringed dyke, with a street on its summit, the river and the flat beyond it, which remains to this day a houseless pasture, at the mercy of freshets. The cross-bar in the H-shaped edifice was one hundred and fifty feet by sixty. The new building, begun near the date of the South Carolinian ordinance of secession, is in size and shape a duplicate of the old, so that the whole forms a double H, with walls uniting the uprights, and all forming a quadrangle of five hundred feet on each side, with several small buildings in its courts—not to speak of storehouses and stables on the north, and several streets

27

of superior tenements for workmen, which run parallel
to the factory, in the rear.

The height of the main building is three stories,
besides an attic, with skylights. The material is for the
most part sandstone, quarried at Portland, a few miles
below on the river. Its color is a rich brown, and the
masonry, being left with many bold projections, after
the style termed rough rustic, is not unlike the lower
story of the Pitti palace, in Florence. Its iron-barred windows are a feature in common with both palaces and
prisons. The front has one hundred and eighty openings
for door or window, and thirty-two skylights. To break
the monotony, the centre is thrown forward some distance
in advance of the wings. Above this centre rises a dome,
—its drum or upright portion white, its bowl bulging in
the Moorish manner, blue, star-spangled, and sustaining a
gilded globe on which a colt is rampant, as it were
triumphant over the world. This prancing steed was
modelled in bronze, and cast in the establishment itself.
Architects would find fault with the front because the
cornice is scarcely perceptible, so that we are reminded
of a face which lacks eyebrows. Several walls also, particularly in the newer portion, are of brick, which is
cheaper, and in the shaky soil can be more easily set on
a firm foundation. But, in these parts, there is a sacrifice
of that grandeur to the eye which is always lost when
there is no individual greatness of material. There is no
making a regiment of grenadiers out of Tom Thumbs.
The handles and hinges on most of the doors are
significantly fashioned in the shape of pistols.

In every apartment there is a hydrant with water for
drinking and ablutions, as well as hose for extinguish-

ing fire. The walls, being of naked masonry, seem also fire-proof. Hence, during the lifetime of Col. Colt, nothing was insured. But insurance appears expedient to the present company, so that they pay more than six thousand dollars a year for policies. Those who insure must pay for protection more than it costs, or whence were the profits of insurance companies, and whence the salaries of their agents? Hence, it is wise for the richest men, and those sure that they are Fortune's minions, to insure themselves. All others will do well to get insured from without.

The apartments, sixteen feet in height, are well ventilated, warmed by hot-air pipes from the engine on the ground floor, and lighted by gas, which, though the operatives are employed no more than ten hours a day, is needed in winter from seven to eight in the morning, and from four to six in the evening. War, says Shakspeare, "makes the night joint laborer with the day." It has here. For the last two years, during more than half of the time, a double force has been employed; half laboring by night. It is worth notice, that the nocturnal workers more often break their tools, and spoil their work, so that they cannot on the whole complete more than half as much as they would by daylight.

In the pistol-shop, the engineer, longest of all employés connected with the concern, neither above his business nor yet below it, lights his fire two hours before work commences, though he says that, in an emergency, he can in a single hour get up steam enough to move all the machines. His pride is to show his wheel, nearly a hundred feet in circumference, and running half a mile in a minute, yet so true and steady that it will not chafe, as it revolves, the board he holds close to its rim or sides. Alongside of his own engine, which he rates at two hundred and forty horse-power, there is another, equal, as he judges, to four hundred and fifty horses, which he can at once fire up in its stead, as a substitute, should his regular motor fail. Such a mishap has thus far never befallen but once, and then, as the beam broke on a Friday—a lucky day for once—it was temporarily repaired by Monday, so that operations were interrupted only a single day. Yet in the nature of things, as well as from the crystallization of iron, such accidents must happen, and any single stoppage of the engine may involve a greater loss than the cost of a substitute, although, like the original, it should be valued at $25,000. In the new shop, there is a third engine of three hundred and fifty horse-power—which, with a fourth "pocket" machine for rolling, make up an aggregate motive force stronger than a thousand horses, smoking, as a cigar, a chimney one hundred and thirty-eight feet high; and while in readiness at any moment to re-enforce the strength of men, like a steed that knows his rider, compliant to the dictates of human skill.

The flooring of the establishment, if all laid in one continuous line, while still as broad as now, would stretch out to the length of a mile and a half, and the ground floors now cover more than three acres. Few palaces then, save the Vatican, are larger, and none ought to be so interesting. By so much as labor is more inno-cent and otherwise better than idleness—by so much as activity that is productive, like the bee's, is paramount to what is unproductive, like the fly's, by so much is an industrial palace a nobler object of rational contemplation than any castle of indolence, luxury, and pride.

Before the rebellion broke out, Col. Colt, foreseeing
that his weapons must ere long be in double demand,
had made all preparations to extend his factory. Accordingly, within ten days after the fall of Sumter, he had
dispatched an agent to England for procuring machinery, such as that for welding musket-barrels. Oh, that
a corresponding foresight of evils to come, and precaution against them, as well as understanding of the times,
so as to know what to do, had characterized our statesmen! Then had half our woes been averted—as thunderbolts by the wand of Franklin.

Both shops, the old and the new, are owned and
managed by one company, with a capital of over a million, and entitled the "Colt's Patent Fire-Arms Manufacturing Company," of which E. K. Root, Esq., who has
been from the outset engaged in the concern, chiefly as an
inventor, or suggesting improvements in the machinery, is
now President. In both shops the machines are to a
great extent similar, while the fixtures on them, and the
tools, are in general very different. The chief end of
the old shop is to make pistols; the new was mainly
erected for turning out muskets, and those fac-similes of
the arms so long manufactured in Springfield, the Government at this crisis being unable from its own armories to supply its own wants. The number of machines
now on hand is fourteen hundred and sixty, and their
cost is estimated by one of the foremen on an average
at $200 apiece. In his judgment the fixtures on machines, and the tools, are at least equally expensive.

In 1855, the Fire-Arms Company—incorporated in

1856—began operations in its present quarters. In glancing at a few of its arrangements, simple but wise, it is apparent that what strikes an outsider as a mighty maze, is not without a plan; the seeming chaos is a cosmos. The association, despairing, as it well might, of time and patience to deal with all individual workmen, who commonly number fifteen hundred, lets out whatever jobs it would get done to a score or two of middle men, or contractors. It provides shops, warms and lights them, furnishes them with machines. Again, it enforces certain general regulations. All workmen, and contractors, for instance, must be at their posts by seven in the morning, and by one in the afternoon, or be excluded for half a day from opportunity to work. No one is allowed to enter, if tardy even so little as an instant. This is greater strictness than that in Enfield, where the British Government allows employés to come in till five minutes past seven, and, if then shut out, to enter at half-past nine. But one must make a stand somewhere, and, fix the moment of exclusion where you will, some laggards will find the doors closed against them. A steamer never leaves New York but we see some belated passenger hurrying down the wharf, when all hurry is in vain. Of course those detained by real sickness are admitted at any time. Laborers may leave at any hour, if they first enter their names with a bookkeeper.

Each contractor, who on the fifteenth of every month is paid up to the first day of the same, engages to fabricate from stock given him, in certain numbers each week, a certain part of an arm, or some part of a part. He then hires his hands by the day, or the piece, as he best can.

Through this system of letting and sub-letting, as well as a gauntlet of inspections, of which it will be in place to speak by and by, a feeling of responsibility and a zeal to be faithful run like the clue of Ariadne through the whole labyrinthine establishment, from top to bottom, keeping all on the right track and working with a will. Each contractor must have his eyes open, or his work will be condemned. The hand-workers must do with their might whatsoever their hands find to do—for it is no holiday affair. Not till they count out the tale agreed on, do they receive their wages. If they break tools, they must smart for it, often sorely. When they spoil a piece, they lose, besides their pains, the value of the stock they waste. This system of mutual checks, works like the law of gravitation in the universe, or Jove's

"Golden adamantine chain,
Whose strong embrace held heaven, and earth, and main."

The result is one of the best illustrations a political economist could desire of the division of labor. In Brewster's Edinburgh Encyclopedia, which was completed no longer ago than 1830, we read: "The manufacture of a gun is performed by the following workmen, viz.: barrel forger, borer, and filer, lock forger and filer, furniture filer, ribber, and breecher, rough stocker, screwer together, polisher and engraver, in all ten different persons, few of whom can execute any branch of the art but one." In Colt's factory—besides machines—more than fifteen times ten men are employed in bringing a single pistol to perfection, a number so great that no language under heaven is copious enough to call them by names descrip-

tive of their duties. The inside of a gun-stock is oiled by
one man, and the outside by another.

Why is the work to be done frittered away among
such a legion? Because all the advantages accruing from
the division of labor, first stated by Adam Smith, and
further developed by Mill, are thus attained.

We first note the increase of *dexterity*, which naturally
grows up not in a scatter-brain, but in him who shuts
himself up in a specialty. We till a garden better than
we can a farm. Even in the simplest operations the
advancement in skill will astonish you. Mark that old
man at work on a screw-driver. With one blow of his
hammer he straightens it, then with another cuts the
trimming at the angle, then brushes it with a file. He
thus turns off three thousand in a day, and as his day has
no more than six hundred minutes, he must finish five
pieces in one minute; while you, outsider, could scarcely
finish one piece in five minutes so as not to bend, or cut,
or file it too much, or too little.

This same operative shows the justness of Adam Smith's
remark,—which some judge ill-grounded,—that division of
labor saves much time that would be lost, were workmen
to change from one kind of work to another. He who
now performs three thousand manipulations on a single
tool, how many would he bring to pass if he undertook
to work upon three thousand tools in a day?

Nor is the third benefit which, according to Smith, is
conferred by division of labor, without illustration at the
armory, namely, that new inventions are thus stimulated.

Here is one specimen. When we would bore a hole
in a block, we commonly turn the gimlet and not the
block, though it seems a matter of indifference which of

the two we turn. In pistol-making, however, it was found by trial far better to turn the thing to be bored, and not the borer.

> Strange such a difference there should be
> 'Twixt tweedledum and tweedledee.

Still, such is the fact. By making cylinders rotate round a chuck, their chambers are hollowed more exactly parallel than is otherwise possible. The machine devised for this purpose is applicable to manifold others. Thus, it best sinks sockets for the rammer and base-pin. It was invented so recently that the patent for it has not yet expired, and it is so valuable that it needs to be jealously guarded against mechanical pirates. In boring after the old fashion a third of the cylinders was spoiled. Not one in a hundred is now. Indeed, contractors hold that none can be, save through the fault of the workmen.

It were impossible to enumerate the improvements in machines, tools, or processes, which in a course of years have been struck out, by men wide awake to save their toil and sweat, as they bear the burden and heat of the day; or by such inventors as Mr. Root, a man who pushes forward to a desired mechanical goal as ardently as a young lover, and as steadily as Ulysses forcing his way homeward to Penelope. Not a few of these improvements are analogous to what has been elsewhere wrought in the art of stocking guns. For this purpose, Blanchard at first devised a single machine, which, since 1820, has given birth to more than a dozen daughters—*Matre pulchra filiæ pulchriores*—who each do more work, and that better and cheaper, than their mother could. No sooner were these machines exported

28

to England, than a workman there contrived another for doing the only thing which the American handiwork had left undone, namely, the boring a hole for the ramrod.

A fourth advantage, not noted by Smith, but laid down in Mill's Political Economy, as springing from divided labors, is the less amount of wages with which the same end may be compassed. If I hire a single man to complete with his own hands an entire gun, I must pay for all the time consumed in constructing it, and so for the simplest processes, the price of a very accomplished handi-craftsman; this is taking a beetle to kill a fly, or climbing over the house to open the gate. Mark the contrast when that arm is made not by one, but by one hundred and fifty-four operatives. No longer need we "cut blocks with a razor." The experts are now better off than any of their class of old, for they earn five dollars a day; while boys, just beginning on the most facile parts of the milling, are content with a tithe of that sum. Besides, those boys are roused to do their best, knowing that all experts were once boys, and hoping to become experts themselves.

We may perhaps add, as a fifth advantage of dividing labor, that it opens a door of occupation to all. If all members in the body politic have not the same office, yet all members have an office, and that the one best suited to them. The hand cannot say to the foot, "I have no need of thee." One man is too ignorant to manage a machine for stocking, but he can oil that machine, or at least sweep away its chips, a service for which two men are paid each a dollar a day. Here then are shallows where a lamb may ford, and depths where the elephant must swim.

"Solomon, seeing the young man Jeroboam, that he was industrious, made him a ruler." So, few contractors but are wise enough, if an underling shows pluck and grit, to try him with a task harder, higher, and more suited to develop him.   Thus, in a single month, he grows skilled in milling; in three months more, in drilling; in a year he can make his own drills; and with study, industry, and brains, after four years longer—just the college curriculum—he may finish an education truly liberal, and graduate as a master modeller.   Thus, in this republic of arts, division of labor

> " Bids each on other for assistance call,
>   Till each man's weakness grows the strength of all."

Who can take no pleasure in an organization which honors all men, and each in proportion to his deserts? It was an old custom in merry England, for the children in a parish to meet at the church on each anniversary of its dedication, and, joining hands, to form a ring round the sacred edifice, that is, to clip it, as the phrase was, in recognition of the blessings of the Gospel.   Something of the feeling which impelled to church-clipping, may fitly glow towards the arms factory in those there employed.   In fact, those who owe it their bread, and hopes of rising in the world, are so numerous, that if drawn up shoulder to shoulder, like soldiers on parade, along the outer wall of the works, they would girdle it with a living and unbroken palisade.   In ancient Greece, a certain visitor at the Olympic games was celebrated as nearest the gods, because his own hands had manufactured his hat, and sandals, and coat, and mantle, and ring, and staff—inasmuch as if the gods had no wants, he was dependent on no one for the supply of his. That ancient was the exact antipode of a modern. The more our mutual dependence, the more our independence. As Dr. Franklin told his fellow-signers of our National Declaration, in '76, we must hang together, or we shall be hanged separate, and one by one, on as high a scaffold as Haman's.

To the question, "How is a revolver made?" the shortest answer which can be given is, "By *revolving*, as we learn to walk by walking." One antique variety of arm was called "wheel-gun," because it had a little steel wheel which, in making a single revolution, struck fire out of a flint that was pressed upon it, as you press the axe you sharpen on a grindstone. With much more reason may we style Colt's arm the wheel-gun. For rotation is its law, and the law of its production. From the huge fly-wheel of the engine, rolling half a mile in a minute, to the tiniest circlet of leather for polishing; in all apartments, even in the attics; beneath all ceilings, in all machines, you are reminded of Ezekiel's vision of wheel on wheel, and wheel in the middle of a wheel. Rotation pervades the products and the producing processes, as in certain monasteries the cross is repeated till you behold it on every bell-pull, the handles and panels of every door, on each floor and ceiling, on every dish and book, so that wherever the inmates move or sit there are crosses to the right of them, to the left of them, before and behind, above and beneath them. In some rooms there are machines by the hundred, with each a dozen wheels, all useless until they revolve. The ramrod wheels, as it forces home the cartridge. The revolving cylinder gave to the revolver its name. The rifle-ball turns round as often as it flies the length of its barrel. Half the parts, being screws, are fixed in position by revolving. On the whole, it may be doubtful whether all the balls ever fired from a revolver make so many revolutions as were made in manufacturing it. Amid so many wheels, all revolving, and that to produce revolvers, what a maddening position were it for that conservative Italian, who had such a horror of political revolutions that he refused to believe in the revolution of the earth! Had that revolution not yet been dreamed of, here was the place for Galileo to have been inspired with convictions of its reality.

The gyrations, many of them, concerning stubborn materials, or delicate processes, no results are apparent to a transient observer, but the universal whirl seems as aimless as the spinning of a top, or the spouting of a fountain. Work of every kind looks like play to children. They never witness an execution without considering it a sort of a joke, and so for weeks afterward they play Jack Ketch. To them the mystery of pistol-making is only a line in Mother Goose: "Here we go round, round, roundy." All outsiders, in their views of mechanical mysteries, are only children of a larger growth.

They look on the processes as hastened, like a juggler's sleight of hand, on purpose to dodge their detecting the secret. Or, they misunderstand their nature as much as one of our statesmen, on his visit to Rome, did that of the Vatican frescoes, when he mistook them for a series of Dan Rice's circus show-bills. But what is in brief the aim of these whirligigs, of this giddy circling, and of the movements, rectilinear or eccentric, into which it is often converted? It is simply to render a few bits of wood, iron, steel, and copper, rough or smooth, hollow or solid, manifold in size, shape, and color; to harmonize them by mutual adaptations, so that they can be compacted as members in one whole body; to produce fac-similes of these patterns outnumbering the myriads in an army, and that in the least time, and at the least cost; as well as to prove, in defiance of all cavils, that these tasks have been smoothly done. This is the purpose of the armory.

How is this purpose accomplished? By thousands of steam-driven machines, which are costly, and whose operations, while seeming slow or roundabout, really attain the ends in view more swiftly, perfectly, and in instances more oft-repeated, than were otherwise possible. Many of these contrivances were invented on the spot. All of them are claimed as American in origin, so that had the geologist Lyell beheld them, he must have pronounced that they, and not Common Schools, were "the most original product of American mind."

As permanent railroads are made by means of provisional railroads, so these machines, and their tools, are made by other machines, which fill a large section of the establishment. For more than one reason, is it expedient to set apart a portion of the works as a machine-shop. Such a shop secures to the company the profits which would otherwise go to machine-makers outside. Again, but for such a department, no small loss would arise through the delay and expense incident to sending to a distance for the repair of the machines and tools which are daily broken. Besides, machines are here made which cannot be bought; the ideas of inventors, a corps of whom are usually on the spot, engaged in suggesting

improvements, are subjected to a fair, but thorough ordeal,
and so far as they come out of this crucible unscathed,
they are actualized, or incarnated in working forms. In
this way, inventions thrive as in a green-house, which
excludes all hurtful influences, while it includes whatever
are auspicious, and keeps its fruit safe from pilferers till
it is ripe. In addition to this, the machine-shop has
been turned to account in shaping the materials used for
enlarging the works, as well as in erecting scores of
dwellings for the workmen. Aside from these utilities,
it is indispensable to afford those test measures, or
gauges, put into the hands of inspectors, which are elaborated with such exceeding tact and pains, that a duplicate set of them for a single musket cannot be made
for less than three thousand dollars. In fine, the more
the machine-shop extends, the more self-contained the
factory becomes, and hence more answerable to the ideal
of its founder.

What, in a general view, is the achievement of the
machines? They soften, and sometimes harden iron and
steel; they beat or stamp them into multitudinous forms,
square, round, multangular, or curved; trim their edges;
plane and turn them as if of soft wood; polish them till
they become mirrors, no less inside than outside; they
bore them through, or tap them; they cut them with
screw threads; they render each piece an exact copy of a
model. These feats are repeated on many pieces at once,
or with such celerity that they can be done a thousand
times in an hour. Some classes of machines which thus
serve as auxiliaries of mechanical toil, or as substitutes
for it, are these—each coming forth with as many variations as a plain song assumes in the voice of the prima

donna at the opera: *drops*, which, raised by steam, so fall
on pattern-moulds as to stamp heated metal, all in a moment, into any shape wished for; *milling*-machines, which
have more than half supplanted the file and cold-chisel;
stocking-machines, mimicking every versatile jerk of the
artisan who used to fabricate gun-stocks, more perfectly
than he can himself, so that his occupation is gone
forever; machines for screw-cutting, barrel-boring, and
broaching; *jigging*-machines, wheels with as many tools
as spokes, and as versatile as the turns in a dance of
"many-twinkling feet;" machines called punches, but
which, aside from stamping out triggers, bolts, and caps
for bullet-moulds, flatten lever-catches, and trim the rough
edges from all varieties of work.

In this factory, as in every other, there are store-
rooms of materials. These of course differ in nothing
from the warehouses from which the iron, steel, copper,
and tin, were brought. They are even rather insignifi-
cant in comparison with the corresponding portions of
some other establishments, because, while elsewhere a
little value is added to much raw material, here much
value is added to a little. The finished product, if a
first-rate pistol, sells for at least fifty times its weight in
unwrought stock, and often for fully its own weight of
silver, so that, were one of the alchemists to rise from
his grave, he would see his dream concerning the trans-
mutation of baser metals to precious ones here realized.

The stock wrought up in the arms factory, iron, steel,
tin, and the files, are imported direct from England—
and that, in a great degree, from the firm of Thomas
Firth and Sons, Sheffield. To be sure, as much steel is
now made in the United States as in any five Euro-

pean establishments, and the American ores, above all, those on Lake Champlain, are regarded as no whit inferior to any in the world. Notwithstanding, the arms company, though not unpatriotic, and in spite of a thirty per cent. duty, and high exchange, cease not to depend altogether on England for raw material. Why is this? Simply, because the English smelt ores, and turn iron to steel, by methods which result in a more homogeneous staple than ours, so that weapons made of it less often fail, when proved and inspected. The truth is, that in fabricating steel, and its mother, iron, young America has made more haste than good speed. Forgetting the time-tried maxim—Make haste slowly, *Festina lente,* he has omitted some laggard process as superfluous. Hence he has fared like the stupid man who, in reporting an anecdote, fearing it will be tedious, brings it to you "just breaking off the point of it, and leaving out the pun." His iron is partly strong and partly broken, like the feet of Nebuchadnezzar's image, which stood tottering, their iron being mixed with miry clay.

It is an old Spanish dogma, that no iron is half so good for gun-barrels as that of old nails, which, driven into the shoes of horses and mules, have been dashed against many a stone. But we can scarcely believe there is any magic in this particular mode of pounding, so haphazard and long drawn out, unless we believe, with Dr. Boyle, that the thigh-bone of a man who has hung on the gallows is better as a medicine than any other. All the force that is applied to horse-nails in making them, and through the kicks of brutes, may clearly be concentrated upon them in a few trip-hammer blows,

29

"turning the accomplishment of many years into an hour-glass."

The real secret of securing good stock is now said to be this. Owing to the law of gravity, and unequal exposure to blast, when as susceptible to impressions as a photographic negative, the lower part of each bar differs in texture from the upper. Let then each piece of iron, and afterward of blistered steel, be broken, and from the fragments let experts, even at this early stage, cull out those nodules which are best adapted for each variety of manufacture. One quality is best for tools, another for arms. Iron-workers can afford to pay more for stock in proportion as it better befits their several specialties. They are confident their handiwork will not fail when put to the proof, if they can secure raw material which suits their needs as exactly as a square man fits a square corner, or as Lincoln's pegs fitted the holes when he sent Butler to New Orleans, and Grant to Vicksburg. Gun-makers agree with theologians, that original sin is the mother of all others.

> "As in a building, if the first lines err,
> If aught impedes the plummet, or the rule
> From its just angles deviate but a hair,
> The total edifice will rise awry."

Yet American steel is improving, and our manufacturers of it are pushing on with such strides that they will soon be no whit behind the British. All hail that day of independence, when they shall excel them in this mother of all arts—the perfection of steel—no less than in locks, yachts, and reapers!

Let us follow a bit of raw material on its pilgrimage. If it be gun-metal, that is, copper alloyed with about ten per cent. of tin, its race is comparatively short. Going into the foundry, it is cast into pistol-guards and back-straps. The moulds are so made that each turns out a dozen guards at once, the molten metal flowing into a sort of tree, with guards as branches. These, when cut off, milled, filed, and polished, pass to the electroplater, from whose batteries and solutions they are received by the burnishers. These brighteners now consist of ten— one man and nine girls. One visitor thought to com-pliment this band by addressing them as Apollo and the nine Muses. But the operatives, being unused to classical allusions, scolded at him for calling them names. For their toil on each pistol, the tariff is three and one-half cents. The females daily complete about thirty sets—the man twice as many. The transfor-mation from dull white to silvery lustre, by dint of mere friction with a steel dipped in soap and water, is astonishing.

Whenever a bar of steel or iron leaves the store-room, it starts on a long journey. It is first heated and cut on an anvil, or with shears, where the cutting is done quicker and better, into pieces short enough for convenient working. Having been again heated, it is, all in a mo-ment, like so much wax or dough, forced into any desired shape on swage-blocks, or anvils cut in patterns. They are thus forced by hammers which, first raised by a screw rotating continuously, are then made to drop by touching a spring. They are hence named "drops." To a non-mechanical eye they seem more properly stamps than hammers, because they finish their work all at one blow —a blow so decisive and ponderous that it recalls the single stroke in the duel between Michael and Satan:—

"The noble stroke uplifted high,
Which might at once determine, and not need repeat,
As not of power at once."

The score of forges, drawn up in a double range, must be in sharp contrast to those of Shakspeare's day, or he would not have represented Hamlet, when confessing himself the chief of sinners, saying his "imaginations were as foul as Vulcan's stithy." Here a lady might venture to walk, with no more exposure of her finery than on a promenade along the streets of Pittsburg. If we may speak in Homeric style, these are a cluster of forges where Vulcan, the very god of iron-work, even when most full of the spirit of Momus, could find nothing to censure in walking from end to end of the files.

The fates awaiting our bit of metal, as it wends its way among the forges, are more various than the possible developments of a thought in the brain of Burns, which might perchance turn out a song, perchance turn out a sermon. According as it is iron or steel, and goes to one forge or another, will it become mainspring or cone-nipple, lock-plate or lever, bullet-mould or pistol-barrel. A pistol-barrel, when stamped by the first drop, has a rude resemblance to a small dumb-bell with flattened heads. The next drop compresses one of its heads into a cylinder as narrow as its neck, leaving the other to form the barrel-shank for admitting the base-pin and ramrod lever. An outsider could not guess what to call a pistol-barrel when first blocked out. It is a bear's cub without form or fashion, which must be licked into shape by many a tongue-stroke of its dam.

The forge department is the special pride of both shops. Its machinery, mainly invented and patented by E. K.

Root, Esq., the present head of the concern, is the most labor-saving, and hence dollar-saving, of all the cunning contrivances here at work. For this kind of labor there was formerly required, at each anvil, a foreman with a helper. Now the helper alone can do more than three times the work of both, and can do it better than lies in the power of any human hands. In blocking out the lock-frame, since it is the most intricate of all members, the superiorities and economies of machinery are most conspicuous.

The apparatus for welding gun-barrels has been introduced since the rebellion broke out. All the barrels previously made, being of steel, needed no welding. In the welding process, flat bars of iron, on passing through the first groove in the rolling-machine, are "crimped," that is, hollowed, into the shape of the capital letter U. Next, they become short and thick as window weights, but are hollow, and with a slit on one side from end to end. When at a white heat, a mandril, or a sort of iron foil, with a guard to screen the workman's hand, is thrust through them, and they are rolled through an eccentric groove between iron wheels, and this squeezing is repeated a dozen times. Each time they are spitted on a thinner foil, and are rolled through a narrower groove. At the close, the side-splitting wounds have healed without a scar, and we behold gun-barrels in length and thickness, though of a bore strangely small. No cavity at all would be left by the elongating and compressing wheels were not the bore filled by the foils, which withstand outside pressure. The two sets of rollers can each furnish forth a hundred barrels in a day.

No words can paint any adequate picture of the gaunt-

let a musket-barrel runs. Yet that its career is long
and hard, will be apparent from a list of some few of
the operations for which names have been coined. The
moment it emerges from the welding-jaws it is plunged
for an instant in water, and, while still red-hot, it is
straightened on the outside, being laid in a steel couch,
which fulfils its mission with a perfection that the proprietors of orthopedic institutes would be glad if they
could equal, or even approach.   Next it is cut off of
the right length, as on the bed of Procrustes. It then
has a cone-nipple, for holding the percussion cap, inserted
into its side. Having been annealed, that it may be softer
for tools, it is nut-bored on a boring-bank, that is, a
rod, being thrust through it, has a nut fastened on its
end, and is gradually pulled back again, so as to enlarge
the hole.   Having next been smooth-bored by a long
gimlet pushed through its whole length, and then
straightened on the inside, it is fastened in a lathe and
turned like a hoe-handle.   Thence it passes to grinding.
This work is done on Nova Scotia stones, the periphe-
ries of which run a mile in a minute.   Such stones,
each more than a foot thick, weighing over forty quin-
tals, and seven feet in diameter, waste away completely,
and perish within two weeks; and the barrels, but for
being kept all the while wet, would shoot out broader
flames than will ever burst from their muzzles.   Then,
sights to help the marksman's aim are brazed upon the
barrel.   After this long preparatory course, comes proving.   The proving-house is a strong cabin in one of
the courts. Thirty-two barrels stand in a stack: one
man puts a tunnel in the muzzles; another pours in powder, from a cup that holds just the proof-charge; a third

follows it with a roll of thick paper. Elongated balls are then put in, and pushed home with a six-pound rammer. The thirty-two are then laid in a separate room, on iron grooves nearly parallel, and pointing into a sand-bank; a train of powder is so drawn along as to unite all the touch-holes, the plank doors are shut, and the whole is exploded by a pistol. The proof-charge is about five times as much powder as one for ordinary service. The wad and ball are also heavier. Not one steel barrel in a thousand fails to bear the test; while those of iron are so much inferior, that as many as five in the thirty-two now and then burst. The place of failure is almost always at the line of welding. That sort of work gives way unless done at the very nick of time, when the metal is at exactly the just medium, neither too hot nor yet too cool. Of six hundred and eighty-one iron barrels lately proved, thirteen burst, and ten were condemned for other faults, so that about one in thirty of them were failures.

It sometimes so falls out, that the proof-charge which will not rend a barrel, will so far weaken the metal that it will fly asunder the next time it is fired, even with a smaller charge. To ascertain whether this is the case, the barrels are tried again, with a charge one-fifth lighter than at first. But it is found that very few that stand the first test prove unequal to the second. As has been said, a thousand steel barrels are proof against the larger charge, where one fails. Hence, the prospect of a flaw in a steel barrel, which has endured the first proof, is too slight to rouse much interest in firing it the second time.

Doubtless it is wise to guard against all possibilities

that a shooter become an involuntary suicide. Still, the
second proof of a steel barrel is not unlike the seeking
the living among the dead, as practiced at Frankfort-on-
the-Main.   In the dead-house there, corpses are laid
in a bed, with a thimble on each finger and toe, so
arranged that the slightest twitch will ring an alarm-bell.
This precaution against premature interment, although
long in use, has never yet detected one single individual
who needed its aid.

An ordinary day's work for two men, aided a trifle
by the United States inspector, will prove about two
hundred and fifty-six pieces—for each of which they
are each paid one cent.

Nor is this all. Our aspirant to the dignity of an
accepted barrel must be "counter-bored" in the breech.
As *counter-bore* is a word you cannot espy either in
Webster unabridged and supplemented, or in Worcester,
some readers may need to be told that counter-boring is
a cut round the head of an orifice, the sides of which are
perpendicular to that orifice, instead, as in counter-*sinking*,
of being beveled down to it.   The would-be barrel is
next "tapped" with shallow holes, like a sugar-tree.
It is then "milled," for squaring or angling the breech.
It is "jigged" on its cone-seat.

The operation which follows is of special nicety, and
is termed *reaming*. It is thrice repeated, and performs
for the inside of a barrel what polishing afterwards
effects for its outside. A wedge of thin paper is here of
power to cause a piece to be accepted or rejected. Nor
is there any thing more beautiful in changeable silk,
or the necks of doves, or in the gay creatures of the
element that in the colors of the rainbow live, and play

in plaited clouds, than in the hues which reaming generates. Our pilgrim tube comes forth from reaming "all glorious within"—but it has still not yet attained, neither is perfect. It is next to be polished. A machine does this service, except that it leaves a few inches at each end to be finished by hand. The value of machines is here brought to view. One machine polishes two hundred barrels a day throughout five-sixths of their length. No less than fourteen men must be busy all the time to complete the single sixth which the machine omits.

Breeching up, deadening or browning, and sometimes rifling, are processes which remain before a barrel can arrive at stocking. In rifling for the United States, only three grooves are cut, and those of uniform width, though growing shallower from breech to muzzle. In those manufactured by the company on its own account, the grooves are seven, of uniform depth, but narrowing from breech to muzzle, as also with a "twist," gaining or graduated. This latter method of rifling is decidedly easier than that required by the Government, and is held by connoisseurs to be in no respect inferior. The apparatus for rifling is American in origin, of late introduction, was mostly made in the arms factory, and is to a considerable extent in use in no other concern.

On the whole, the barrel machinery sets it careering on such a many-twinkling dance, now advancing, now retreating, as Cowper had in mind when he thus wrote:—
"I have read before, of a room with a floor, laid upon springs, and such like things, with so much art in every part, that when you went in, you were forced to begin a minuet pace, with an air and a grace, swimming about, now in and now out, with a deal of state, in a figure

30

of eight, without pipe or string, or any such thing, but making you dance, and as you advance, 'twill keep you still, though against your will, dancing away, alert and gay, with bustling ado, till madam and you are quite worn out in *jigging* about."

To those who judge by the sight of their eyes, there is no marvel in musket-making comparable to the half-reasoning machines which fashion the stocks. These are now sixteen. Receiving a rough and crooked rail, they send it flying through a course of sawing, and centering, and spotting, and first-turning, and barrel-grooving, and profiling, and butt-plating, and letting in the lock, and banding, and working between the bands, and ramrod-grooving, and bedding guard and ramrod spring, and second-turning, and boring and tapping. One is at a loss to decide whether each successive process subtracts more from the weight, or adds more to beauty and adaptations.

These manipulations seem as cruel as the joint-cracking and other terrors of a Turkish bath; and yet, all the while, they are as safe, salutary, and suppling, as well as fatal to all clogs on the patient they take in hand.

These stocking engines are a Yankee notion. They were invented by Thomas Blanchard, who was born at Sutton, Worcester County, Mass., in 1788.

The history and nature of his invention deserve special notice. Mr. Blanchard had devised a lathe to turn musket-barrels with a uniform external finish. Knowledge of this invention came to the superintendent of the Springfield armory, who contracted for one of the machines. When it was put in operation, one of the workmen remarked that his own work of grinding the barrels was done away with. Another, employed upon the wooden stocks, which were then all made by hand, said that Blanchard could not spoil his job, as he could not make a machine to turn a gunstock. Blanchard answered that he was not sure, but he would think about it; and as he was driving home through the town of Brimfield, the idea of his lathe for turning irregular forms suddenly struck him. In his emotion he shouted, "I have got it! I have got it!"

The principle of this machine is, that forms are turned by a pattern having the exact shape of the object to be produced, and which in every part of it is successively brought in contact with a small friction wheel; this wheel precisely regulates the motion of chisels arranged upon a cutting-wheel acting upon the rough block, so that as the friction wheel successively traverses every portion of the rotating pattern, the cutting-wheel pares off the superabundant wood from end to end of the block, leaving a precise fac-simile of the model.

The whole sixteen, which, within a generation, have grown from a single germ, were ordered by the British Government for their establishment at Enfield, and the very man, Mr. Oramel Clark, who now manages them here, set them up there, and managed them for more than four years. The only fraction which American machines omitted, some English artisan has added a machine to do. Seventeen engines, then, construct a gunstock, perfect in all essential parts. As they leave it, it is in a shape—as soon as you whittle two or three scarcely perceptible notches for the angles of the barrel and the curve of the breech-pin, a trifle, for doing which no contractor would pay you more than nine mills—all ready to receive lock, barrel, bands, butt-plate, guard, ramrod, and hence to be used in actual warfare.

The stock does not in fact, however, thus receive its mountings. On the other hand, it is spoke-shaved, filed, and sand-papered, with much ado about a very little, for smoothness, fancy, and Government exactions. Here—let it be noted—is a pre-eminent proof how economical machinery approves itself in contrast with hand-labor. The single hand operation, which is a non-essential, both employs twice more men, and costs twice more money, than all the sixteen machine operations, every one of which is essential. The machine is a factotum; the man a do-nothing.

Machine stocks are better than handicraftsmen could construct in at least two points; they interchange at will with each other, and they are less liable to be condemned by United States inspectors, as too large or too small. In the Government ideal, the stock is as light as possible consistently with strength, and as strong as possible consistently with lightness. All deviations from this stand-ard, this bundle of compromises, however trivial, are therefore to be blamed as sins against either lightness or strength. If too heavy, the musket wears out the soldier, as the last peppercorn breaks a camel's back; if too light, its recoil lames his shoulder—like the Hudibrastic gun, which kicked wide and knocked its owner over.

No operation in gun-stocking is more magical than cutting the hole for letting in the lock. A hexagonal frame, No. 1348, is so hung as to turn round above the stock. From each of its corners, as it halts an instant, a tool springs down fiercely, and seems to be tearing the stock in pieces. All the while a blast from two brass pipes blows away the fragments. At the end of forty-five seconds, we behold, hollowed out of the solid wood, a polished cavity of five different depths, and so cunningly fashioned that the most shapeless of all contrivances, a gun-lock, rests and plays in it as in the niche it was ordained to fill.

The black walnut of which stocks are made, used at Springfield to be seasoned for four years. That process is here, by kiln-drying, shortened to but little more than that number of weeks.

On a first ramble through the arms factory, few persons, not of a mechanical turn, scrutinize more than that small fraction of its routine which we have now passed in review. But in every thoughtful mind, curiosity regarding cunning arts is an appetite which grows by what it feeds on. Every member in the arm has a history worth tracing did our limits allow, or could it be made as intelligible in description as it is when surveyed by the eye.

On some visits, we may take a view not so much of any thing in particular as of the magnitude of the whole. This is so great that, in 1858, the Secretary of War described it, though not yet half its present size, "as having risen to the dignity of a great national work, as well as superior in machinery and extent to either of the national armories of the United States." During the last ten years, it has turned out more weapons than were made or purchased by the British Government in the ten years between 1844 and 1854.

On a general survey of the *tout ensemble*, we shall linger longest in the armories proper—that is, in the second stories of the old and of the new edifice. Here

are two rooms, each five hundred feet by sixty, each
filled with from four to six rows of machines, and
around the walls vises enough to hold all the virtues
of the handiwork.   Toward the river and fields beyond
there is a charming outlook—but as one man sometimes
tends six machines, there is small opportunity to look
out.   All work so intensely that they salute no man by
the way.   From choice they keep that silence which is
enforced at Wethersfield and Sing Sing.

"Within his mouth each doth enjail his tongue."

Near evening, however, we fall in with workmen with
their sinewy vigor all exhausted, and so ready to talk,
lest through overdoing to-day, they become unequal to
the task of to-morrow.

We come then into chat with a filer.   When a jigging
machine has rudely scooped out the shank of a pistol-barrel, it passes into the hands of the first filer, who is paid
seven cents for each of those tubes he faithfully rubs
something more into shape.   His work is hard, and his
pay proportionate.   From him, after sundry other manipulations, it comes to the second filer.   He uses seven files,
costing in the aggregate $2.29, as bought of the company
at wholesale, and though he cards them clean a dozen
times a day, he wears out two or three sets every month.
He uses them all on the shank, or hook, or sight, and
makes on an average twenty strokes with each.   For
these one hundred and forty strokes he is paid but two
and a half cents; yet, as one hundred and fifty barrels
daily pass through his hands, his wages are good.   He
stamps his initial on each, lest the inspector send back
to him the bad work of another, so that he would suf-

fer for another's fault. He has not been long at the business, and would not be able to do his work, but that he was brought up a machinist. When he screws up a barrel in a vise, why do not the jaws scratch it? They are faced with leather stuck on by beeswax. Through fear of damaging the work, most hammers are made from a soft alloy of tin and lead, called Babbit metal.

Few can walk through the armory without observing its adroit economies. In the forging department, for instance, the first thing we see on our right is a ponderous frame called Dick's shears, which cut off from bars of steel, according to a gauge, the exact quantum needed for each particular article. Not a little of raw material is thus saved. Besides, each piece, having nothing superfluous, comes out of the die so clean as to be easily trimmed. All the scraps, however small, as well as borings, filings, and shavings, are gathered up and sold. But as this metallic shoddy sells low, yet is as much better than the original article as other shoddy is worse, apparatus is just now preparing for working it over into bars again, in the arms factory itself. In most cases, the scraps as they fall from the machines are drenched with oil. This union of oil and iron—as Charles Lamb said of mixing brandy and water—spoils two good things. Accordingly, the metallic shavings being collected in a reservoir with a bottom full of holes, they are steamed till the oil is drained off. The steel threads scraped off in rifling resemble wool, and are squeezed in a sort of cheese-press till cleaned of oil. Whenever it is possible, soda-water, being cheaper, is substituted for oil as a cooler and lubricator. There are ten tanks of such water.

But for these and such like devices, the expense of oil

would be doubled. How much is thus saved we cannot appreciate, till we count the thirteen cisterns in one row stored with oil, and ascertain that the consumption of it, after all, last year amounted to more than eleven thousand gallons [11,377 1-4 gallons].

Again, for forging an iron barrel, it is necessary to begin with a plate so heavy that a tip, several inches long, is usually cut off from one end after boring. These tips used to be sold at two cents a pound. But one, or at most two, such tips will make a lock-plate, the iron for which costs eight times as much. Therefore barrel-tips are now kept, and are forged into lock-plates.

Again, the stocks of many pistols are fabricated of ivory. More than half of each tusk would seem to be wasted while being brought into the fitting shape. Ivory chips, however, and sawdust are not only a better fertilizer than guano; they are a daily necessity in the shop itself, for case-hardening every article which is wrought out of iron.

Yet once more. Cotton waste is indispensable at most machines for wiping them, and the workmen's hands. According to dictionaries, this article is so called "because it is of little or no account or value." Here, it is not only preserved with care, but, when dirty, is washed and used again and again.

On the whole, the more we look, the more on every side do we espy economies—those littles which make a mickle—and which could never have been thought of but by a Yankee who had full faith in the German superstition, that whatever we leave uneaten on our plate— or unutilized anywhere—is a sacrifice to Satan.

The parts in a rifled musket are eighty-four—and several of them, for describing their genesis, would each require as much space as has been given to the barrel or the stock. At every turn in the workshops we stumble on something new or curious. It is long a mystery how steel can cut steel, and that to all appearance as easily as the softest wood. In such cases the cutter is of harder temper, and its hardness is re-enforced by a diamond-shaped point. In making test gauges, where hard steel is to be cut or planed, the tool takes off at once only a very narrow as well as thin shaving. If we watch this process, we shall sometimes find that twenty-seven strokes are required to plane a surface one-eighth of an inch wide.

Other tools, as punches, are greatly larger than what they act on, and so, striking on a steel plate, cut out navy triggers as rapidly as a baker stamps out crackers in dough.

It is no wonder that Col. Colt invented more than one sort of lubricator, to one who marks how much oil is consumed in the construction of arms. Some of the tools could not be tempered aright unless dipped in oil. Of course there is oil on all wheels, whether on engines below, or on machines above. How it would clog all arms-making here, did we hold with Brazilians, that it is impossible to drive away devils without leaving all wheels ungreased and creaking! There is oil on all bright iron, lest it rust; oil on all tools, and the metals they cut, to guard against heat, and the friction, technically called "grinding," of borings, shavings, and parings. On the inside of a gunstock there is sperm oil to soften it; on its outside there is linseed oil to harden it. In manifold delicate manœuvres, it is clear that the

31

oil—so trifling in quantity, and so mild in its action as
not to be noticed—makes the difference not merely
between facility and difficulty, but between facility and
failure.

On many machines we see brushes, self-moved, which
remove all rubbish, so that it cannot obstruct the tools.
Most machines are automatic, or self-guided, stop when
they have finished their task, and some of them are said
to do so whenever any thing is deranged, so that if
they go ahead they must go wrong. They must be sure
they are right, or they will not go ahead.

Including screws, there are twenty-eight pieces in a
pistol, and three times as many in a regulation musket.
Most of them, viewed singly, seem useless. Indeed, they
are so. Their value is apparent only when they are fitly
joined together in one organic whole compacted of many
parts, all mutually related—*E' pluribus unum.* Their
strength, like that of the old man's fagot in Æsop, lies
in the band that unites them. It will move our special
wonder in the assembling-room, where dead fractions
are vivified, to see how readily, after a slight touch with
a file, and more often without any fitting, pieces taken
at random from heaps of the same sort, fit into each
other. It reminds of the reunions at the resurrection, or
of Solomon's temple, "the house which, when it was in
building, was built of stone made ready before it was
brought thither, so that there was neither hammer, nor
axe, nor any tool of iron, heard in the house while it
was in building." How is such an "assembling" possible? Because all the bits of the same name have been
formed in the same die, or cast in the same mould, and
measured with the same exactitude, by the self-same

gauges. As long ago as the Mexican war, so exquisite was the adaptation of member to member, that three-fourths of the broken arms picked up on a field of battle, interchanged so as to furnish forth serviceable weapons again. This matter of interchange deserves the more notice, because it has been both exaggerated and underrated.

It is not easy to exaggerate the readiness of interchange in the parts of a musket. So we shall judge when we have stood by the assembler, seen him pick up the pieces at hap-hazard, clap them together, and fasten them by eleven screws, and all within three minutes. That the members composing a pistol may each find in another its counterpart, some slight manipulations may be needed. Yet the system of correspondences has never been carried further than here—no, not in Swedenborg's Heavenly Arcana. It is no rhetorical flourish to say, that parts into parts reciprocally shoot as perfectly as Leibnitz could imagine them in his doctrine of Pre-established Harmony. Accordingly, orders come every day for some single piece—barrel, hammer, trigger, or lock-frame—to replace one that has failed. Such orders are given, and filled out, with well-grounded confidence that the new-comer will at once find itself at home in its new position, though a thousand leagues away, and in an arm fabricated years ago. There is then a practical advantage resulting from rendering each element in a weapon a fac-simile of the analogous element in every other of its class. Again, if a model piece is the *beau idéal* in size and form, all deviations from either—however infinitesimal—are so far faulty and reprehensible. The metallic portions are of course of the same weight, so that

when an artisan has finished a day's work of several thousand bolts, for instance, and is paid so much a hundred
for them, he has only to count a single hundred, and
then ascertain by scales how many times as much as
that hundred his whole quantity weighs.

It will still be asked, "If such permutations and combinations be possible, why is a number stamped upon
each pistol six times over?" Partly because pieces which
have been once matched in the assembling-room, unite
with rather less ado than others; partly also to show
how many have been made, and facilitate the keeping of
accounts; and in great part to aid in identifying stolen
goods. This expedient to stop a thief is analogous to
what is termed "rogue's yarn," that is, yarn of a different
twist and color, which is inserted in all the cordage of
the British navy, to identify it if stolen. The boys who
bring workmen their dinner, now and then snap up a
screw, or bolt, or trigger, as an unconsidered trifle; then
go on to pilfer a stock, or cylinder, or barrel. When
they have secured all the eight and twenty component
parts, they assemble them in a pistol. But such a pic-
nic arm, wherever seen, betrays itself by lacking the
number, which is not stamped in the factory until all the
parts are united. More than one theft having been thus
detected, such thieving has hence become rare. These
pistol-stealers would have eluded the detectives, could
they have erased the numbers from half the revolvers in
Hartford, repeating the ruse of a certain French page.
This scapegrace, when caught in the chamber of a maid
of honor, had managed to escape unrecognized, but with
the loss of his badge. Next morning, all the pages were
called before the major-domo, to see which of them had

no badge. None of them had any, for the guilty one had gone to the dormitory of his companions, and, while they were asleep, had abstracted all their badges. Seldom has necessity been mother of so artful a dodge. Again, when it was once inquired in court whether certain cases of arms had been stolen from the United States, it was proved that they must have been, by the numbers on them corresponding to those recorded on the books of the company as sold to the Government.

Before being "assembled," or combined in one, each individual portion of each arm is inspected. In Japan, all cross-eyed men inherit as their birthright the office of spies, for all who have that peculiarity of vision are viewed as "fellows by the hand of nature marked, quoted, and signed" to do a Paul Pry's work. In like manner, inspectors ought to be selected from the most thorough-paced grumblers,—or what Hood calls "hedgehogs rolled up the wrong way,"—so that their duties and inclinations may all point in the same direction. Their duty is to cavil on the ninth part of a hair. We take up a tumbler which has already stood the ordeal six times, and drop it into a gauge. It slides in like the noiseless flow of oil from a flask, and sets so tight that it will not rattle. Yet it is rejected. Why so? Simply because on one side it fails to fill the gauge-socket so far that you can thrust in a corner of a bit of silk paper. When the inspector tries a barrel, his gauge must sink in without friction, yet touch it at every point, so that the air compressed below will hold up the bar of steel, and, if you force it down, will toss it up again. He has a block of lead on which he hammers triggers set up edgewise, to see if they will break. Lock-frames he knocks against

this block, and, if they ring dull and dumpish, condemns
them as lacking a tight base-pin. Cone-nipples he twirls
on an awl, and throws them out if it goes through
beyond a certain point, or does not reach it. If any chamber in a cylinder falls short a hair's breadth of the
standard depth, he sends it to be bored again. Occasionally, a scar in the lock-frame which mars not its
efficiency, is winked at by the company's officials, but it
finds no mercy among those sent hither from Washington.
Now and then, those dressed in a little brief authority
play such fantastic tricks that they are compared to that
stupid ship-captain who boxed the compass for a week
round a shoal which was at length found to be an imaginary danger,—namely, a fly-blow on his chart. Here
the inspector has condemned a revolver which to our
eyes is perfect. Why is it not? Because one shaving
too much was pared off on one side of the back-strap.
The inspector has also cut an ugly notch in its stock,
and if we ask him why, answers, "That I may know it
again. If I did not nick it they would send it back to
me once more, and when the defect is very slight and
microscopical, I might accept to-morrow what I have
rejected to-day, and should then be laughed at for incon-
sistency." How often, with accidental variations, men
repeat the experience of those riotous youth, who, when
King Philip gave judgment against them, appealed, and
when he asked "To whom?" said, "From Philip drunk
to Philip sober."

But, trifling apart, the United States inspectors ought
to be hypercritical, lest by any possibility an arm pass
muster with them, and yet fail when it cannot be repaired, or in the heat of battle, when its failure is likely

to cost a life. Hence, few officials deserve their salaries better than the six-and-twenty professional fault-finders whom the Government here maintains. With reason also are this corps transferred, at short and uncertain intervals, from armory to armory, that they may not be open to bribes, or worth bribing. Nor need Colt's arms shrink from any test, however rigid, for they have obtained from the Secretary of War's censors as good a report as those of any other manufactory, whether private or national. Indeed, four muskets here made have been accepted by the Government as models, and sent as standards to contractors at other armories in four different States.

Yet have not our authorities, after all, been penny wise but pound foolish? While so scrupulous about muskets, they have been negligent about the men who shall handle our armies. Cadets have been admitted at West Point, because franked thither by members of Congress, or because the pets of higher functionaries, not because they have given tokens of capacity to excel in military science. Had that school been open to those only who, in the fair field of competitive examinations, could prove themselves worthiest, how much more ably would our forces now be officered! How much more grateful, hopeful, and serviceable would be the task of its instructional corps! Had its doors opened only to receptive minds, its professors would never have been expected to carve marble statues out of snow, or, according to the proverb, to make a silk purse out of a sow's ear.

Such is a meagre sketch of some characteristics of what may be styled the cornucopia of Mars—the establishment in which fire-arms of the best model were first fabricated on a grand scale in the United States by a private com-

pany. Col. Colt lies buried near his house, in a thicket
of evergreens, and beneath a plain marble slab. But his
simple tomb is in full view of the arms factory, which is
his true monument. Before his death, and through his
energetic genius, it had become substantially what it is
to-day. The volleys in the proving-house daily sound his
requiem. *Si monumentum requiris, circumspice!*

DESTRUCTION OF THE ARMORY BY FIRE.

For one who has written concerning the Armory with
such affectionate minuteness as in the foregoing description—of its rise, progress, and multifarious departments—
it is a sad task to chronicle its downfall—plunged in an
ccean of fire. Yet thus it must be.

On the 5th of February, 1864, the writer of this paper
was railroading from the West to Milwaukee. He met a
friend from Marquette, on Lake Superior, and was asked,
"Have you heard the news?" and on answering "No,"
was told that Colt's Pistol Factory was destroyed. He felt
the pang which pierces in a personal bereavement, a pang
which was renewed the next summer when he rambled
about the ruins, stones fire-blackened and ready to crum-
ble through heat, machines tortured into strange shapes,
and walls tottering to their fall among weeds, with
which nature was already beginning to mantle the desolation. The scene was such as we associate not so much
with the new world as with the old, where stones, themselves to ruin grown, are gray and death-like old.

On that February morning the Armory labors commenced at the usual hour of seven, and all was well for
an hour. Very soon after eight o'clock, some of those
engaged in the upper story, noticing smoke near its ceiling, gave an alarm and ran up stairs. Others followed,
with hose from the hydrant, which they pointed at a black
vapor that was streaming out of the drying-room; but
the water, from some unexplained cause, came not until
too late. The drops, which at this moment could
have extinguished all, would have been worth their

32

weight in gold. A little fire is quickly trodden out, which, being suffered, rivers cannot quench. No water came, but fire did. Thus long it had smouldered among the patterns in the drying-room. It now burst forth, drove the men from the attic, and caught the roof beneath the dome. The attic floor was soon burned through, so that coals fell on that of the third story, which, composed of yellow pine, and soaked in oil from the drippings of machinery, at once spread the fire every way faster than a man could walk. Thus it fared with the second story, and thus with the lowest. Within one hour after the first smoke was discovered, the roof, with its dome and rampant colt, its supports having been consumed, fell in, re-enforcing the flames that were already all-conquering beneath. Each floor was an acre of flame— all machinery covered with oil, all wood saturated with it, ten thousand gallons in reservoirs, besides other combustibles, augmented the fury of the conflagration, already uncontrollable.

Men did what they could. The steam gong, which had been heard fifteen miles, was heard through the city sooner than any cry or bell. The fire companies with their steamers were at once in motion. But the water in cisterns was scanty, and it must be brought more than seven hundred feet from the river; besides this, the hose burst more than once. Yet, had there been none of these untoward accidents, it is by no means certain that the best possible play of the engines could have stopped the raging element, after it had gained full headway, sooner than it was in fact arrested, namely, midway between the old and the new factories. The combustion was most intense and rapid, till there was nothing more in the old factory to burn. Thus perished the whole original erection of Colt, the chief basis of his fortune and fame.

The further progress of the flames was arrested mainly through the herculean efforts of the armorers themselves. An interested observer of the scene writes : " As they saw wall after wall weakened by the heat, and falling beneath the play of the engines, they rose in their strength to do battle. But for them, doubtless the whole, instead of the half, of that noble pile of buildings, the mechanics' pride, had been laid low. Each taking the bucket provided for him, to remove from face and hands the marks of toil before going to his home at noon and night, and forming lines from the hydrants to the roof, in the part known as the New Armory, they determined to save it. Some upon the slate roof, so near to the burning mass that it seemed impossible humanity could endure it longer—amid the stifling smoke, and the groans of the steam gong, almost human in its wailing—beside the falling walls of their home for many a year—did these brave men labor with the might that only affection can give; they loved their chief, who had gone from them forever, and they willed that their hands should save his monument from destruction. Thus, one line passing down the empty buckets as the full ones came up—at times almost compelled, by the fearful heat of the atmosphere about them, and the almost burning slate roof under their feet, to abandon all to the fierce fire— they at length drove it back inch by inch, until at length, though worn and weary, they came down victorious—triumphant. God bless those brave, true men, now and for aye!" Their efforts, by saving the New Armory, enabled the company in three days to go on with the completion

of their contract with the Government, without asking
for any extension of time.

The office might have been saved, had the bridge
uniting it to the main edifice been cut away in season.
When a tardy effort was begun in this direction, it was
thwarted by some one's shouting from the crowd that
there was a ton of powder in the cellar. Owing to the
machine-drawings and other valuables in the office, its
destruction, though it was not a large building, swelled
the loss by some hundreds of thousands. The lower
story of it was in fact saved; but the main struggle of
the engine-men was to rescue the new building, and in
this they succeeded. One loss, much deplored, was that
of certain unique machines—tricks to show the extent of
human brain, the models of which also perished—for executing certain delicate and important tasks most quickly,
perfectly, and cheaply. Some costly things went to ruin
because the windows were grated, the workmen locked
in, and a principal door for a time impossible to open.
Among these, some (as the screw machines) were the
only specimens of their class this side the Atlantic, and
not to be at once replaced. The value of the machines
lost was estimated at $800,000, that of the stock at half
that sum, and the insurance covered about one-third of
the amount destroyed. One man was burned to death;
several others, hanging out of high windows by their
hands, could not reach the ladders with their feet. They
must have perished, had not their friends bethought themselves and been strong enough to lift those ladders up to
their feet. More than eight hundred men were in one
moment thrown out of work, while the injury to the
Arms Company and to the country was aggravated by the

check given to pistol-making in the very crisis of a war when they were more than ever needed. In this regard a significant anecdote is related respecting a speculator in New York, who, the minute he saw the telegram that the Armory was on fire, hurried to the largest pistol dealer there, and bought up his whole stock before that dealer learned the news.

The greatest fire-plague that has ever smitten Hartford cannot be forgotten by any who witnessed it; it is conceivable only by those whose own eyes have beheld a conflagration equally vast. Instead, therefore, of declaiming upon the catastrophe in stilted prose, we shall better recall it, or paint it, in the words of the fire-scene in Schiller's Song of the Bell, a portraiture which genius has so drawn that, as long as fires burn, men will never cease to recognize its truthfulness, thus verifying the sage maxim of Aristotle, that poetry utters universal truth, and hence is more truthful than history.

> "An instrument of good is fire,
> With man to watch and tame its ire;
> And all he forges, all he makes,
> The virtue of the flame partakes;
> But frightfully it rages, when
> It breaks away from every chain,
> And sweeps along its own wild way,
> Child of nature, stern and free.
> Woe, if once, with deafening roar,
>   Naught its fury to withstand,
> Through the peopled streets it pour,
>   Hurling wide the deadly brand!
> Eager the elements devour
>   Every work of human hand.
>
> Hark! what tumult now
> Rends the sky!
> Lo! the smoke up-rolling high!
> Flickering mount the fiery shafts;—
> Where the wind its wild wave wafts,

Onward through the street's long course
Rolls the flame with gathering force;
As in an oven's jaws, the air
Heated glows with ruddy glare;
Falling fast the rafters shatter,
Pillars crash and windows clatter.
Through the air in graceful bows
Shoots the watery stream on high.
Fierce the howling tempest grows;
Swiftly borne upon the blast,
Rides the flame, devouring fast;
Roaring, crackling, it consumes
All the crowded armory rooms;
All the rafters blaze on high;
And, as if 't would tear away
Earth's foundations in its flight,
On it mounts to heaven's height,
Giant-tall!
Hope hath all
Man forsaken; helpless now
He to heavenly might must bow,
Idly musing o'er his fall,
Wondering at his work laid low."

The resourceful spirit of the Fire-Arms Company was shown in that their workmen in the unburned half of their establishment, on the second secular day after the conflagration, were at their labors again as if nothing had happened. It is further manifest in their now rebuilding just on its old foundations, and more incombustibly, whatever fire had thrown down. Thus is also fulfilled a prophecy in which, writing of the Armory three years ago—then in its pristine perfection—we indulged, namely: "No matter though his manufactory be burned to ashes, and not one stone be left on another, his invention would rise as from the pyre of the Phœnix, and clothe itself in a new, perhaps a more noble embodiment."

Regarding the origin of the fire, opinions will perhaps never coincide. Some traced it to rebel emissaries, such

as burned Western steamers and New York hotels; others
were convinced that cotton waste, used in scouring attic
machinery, had been left in the drying-room; others
judged the heat of the steam there had become so great
as to ignite the dry wood.

Had Col. Colt survived until the day of the fire, insurance men tell us that the loss of the company would
have been four hundred thousand dollars greater, inasmuch as he never insured his property. On the other
hand, had he still lived, the fire might not have
occurred. His quick-eyed, persistent vigilance, for which
nothing was either too vast or too minute, might have
kept the water-pipe in working order, or deterred the
scourer from leaving his waste in a fatal spot, or have
hurried away the dried wood before it began to burn.
After all, what might have been is beyond our judging.

> "There's a divinity that shapes our ends,
> Rough-hew them how we will."

Well is it said, "Higher spirits may discern the minute
fibres of an event stretching through all time and space,
and hanging on the remotest limits of the past and of
the future, of the near and of the distant, where man sees
no more than a point, isolated and unconnected. The
changes, then, which would flow from the change of any
antecedent, however trivial, who shall not confess it beyond his conjecture?"

All sympathized with the author of the following lines:—

ON THE BURNING OF THE ARMORY, ERECTED BY THE LATE COLONEL COLT.

O storm of fire!—O surging flame!
That from the depth of mystery came
To wreck, with demon rage and roar,
The proudest pile our region bore!
Through casement, tower, and battlement
Hot, hissing serpent-tongues were sent—
Down came the arches strong and bold,
Down sank the dome of blue and gold,
Mid clouds of smoke and shrieks of steam,
And molten metals' dazzling stream—
Nor slack'd the wrath till all around
A Balbec-ruin strewed the ground.

O storm of fire!—O leaping flame!—
Rejoicing in thy work of shame,
And heralding the ruthless deed
By telegraph, that all might read
On reddening sky—for leagues away—
Loss!—loss!—and grief! and tyrant sway;—
While throngs that drew from year to year
Subsistence for their households here—
Earn'd 'neath yon roof their children's bread—
View desolation round them spread,
And raise that spirit-wail of woe
Which none save those who share can know.

O storm of fire!—O reckless flame!—
Intent to feed upon *his* fame
Whose genius as by magic wrought
Here, in a wilderness of thought,
And with the vast machinery twined
The impulses of a ruling mind—
That fame shall glow in quenchless trust,
When thou but ashes art, and dust—
That fame on Phœnix wing shall soar—
Yon lofty fane arise once more—
And gathering to its shrine be seen
The sons of toil with hopeful mien—
Repeating still through future days
Their cherished benefactor's praise.

                                        L. H. SIGOURNEY.

HARTFORD, *Friday, Feb. 5th*, 1864.

ELISHA K. ROOT.

NEXT to the death of the great inventor of the Colt Revolver, the Patent Fire-Arms Company, as well as this whole community in the field of mechanical invention, could sustain no greater loss than in the death of Elisha K. Root, on the 5th of July, 1865. This eminent machinist and excellent man was born in Belchertown, Hampshire County, in the State of Massachusetts, May 5th, 1808. His education, in the ordinary acceptation of that term, was confined to a diligent improvement of such opportunities as a common district school could afford, for four months in the year, until he was fifteen years old, when he became apprentice in a machine-shop in Ware. But his mind was stimulated by being early associated with other minds in a cotton factory, where he worked as a "bobbin boy" for eight months in the year, from the time he was ten years old until he began his apprenticeship. Here, as well as in his subsequent career, he was familiar with great operations, and surrounded every moment with striking illustrations of the triumph of mind over matter. Every thing with which he had to do was an eloquent witness to the value of education—to the power of mind in devising and improving machinery, and in wielding the gigantic forces of nature to accomplish splendid pecuniary results, as well as moral good for mankind. In his peculiar field of work his thoughtful mind was trained to habits of observation and reflection, and his gift of invention was developed into a beneficent power. After serving his apprenticeship in the

33

machine-shop in Ware, he worked as a machinist in various factory villages, and among them at Stafford, Connecticut, and at Chicopee Falls, Massachusetts, returning occasionally and for brief periods to Ware. On one of these visits he saw, for the first time, Samuel Colt, astonishing boys and men by an experiment in blowing up a raft on Mill Pond by a torpedo, or some preparation of powder. In this, as well as in his early experiments with the pistol, "the boy was father to the man."

In 1832 Mr. Root removed to Connecticut, and became connected with the Collins Company, in Collinsville. The originator of that successful company writes: " He came here and offered his services in 1832, a young man about twenty-five years of age, without any recommendation, except a remarkable head and eye, indicating a man of more than usual mental ability. He called himself a machinist, and commenced work at a turning-lathe in our repair-shop. It was not long before his superiority became manifest, and he was appointed overseer of that shop, and in a few years was virtually overseer of all the shops, though not appointed superintendent by our directors until 1845. He invented several useful machines for facilitating our work, some of which were patented. He never manifested anxiety to obtain large compensation for his services, but was content to bide his time. In 1845 he was offered the situation of 'master armorer' in the United States Armory in Springfield, and about the same time he received two offers from large manufacturing concerns in Massachusetts, with very liberal compensation, which resulted in our giving him increased pay to remain with us. In 1849 Col. Colt made him very liberal offers, more than we could afford to pay, but he was satisfied with his compensation here, and would probably have remained, if I had not advised him to accept this offer as a matter of duty to himself and his family. He was here with me seventeen years, and I knew him intimately. He was not only a superior mechanic, with great inventive faculties, but he was a man of excellent judgment, and great caution and prudence, which cannot be said of inventors generally. He was also a deep thinker on most subjects that interest men of science and thought. He was a man of great simplicity and purity of character, a very modest, unassuming man, and yet very decided and firm. He was a very conscientious man, so that all who had business with him were impressed with his strict honesty and integrity. He was a man of liberality, in every sense of the word. He was averse to display of any kind, and despised all sham. He was a very superior man, and he had few equals. Such a man is a great loss to any community. Any thing I could say would be deemed faint praise by those here who knew and loved him."

While at Collinsville he invented and patented several machines which greatly facilitated and perfected the manufacture of the axe, by which the Collins axe got and kept possession of the American market. One of his patents covered an entirely new process for punching out the eyes of axes, instead of forming them by welding; and another, by bringing the axe to an edge by chipping instead of the slower process of grinding, at once economized its construction, and obviated a deleterious result to the lungs of the operatives.

After joining Col. Colt as superintendent of the manu-

facture of his arm at Hartford, he devoted all his efforts
to perfecting the machinery by which alone an effective
weapon could be made with economy; and in this direction he achieved results which no one was so prompt to
perceive, adopt, and reward, as the great inventor of the
Revolver.   Besides introducing, from year to year, improvements in the details of almost every process in the
construction and arrangement of the parts for which he
took out no patents, he devised in 1853, in the drop-
hammer, a new mode of forging the parts of fire-
arms, which was patented, and which has been widely
introduced into all forging shops in this country and in
Europe.   In 1854 he procured a patent for an improved
machine for boring the chambers in the cylinders of
revolvers, which completely revolutionized the manufacture
of the article, by the accuracy and rapidity with which
the work was done.   In the same year he patented a
compound rifling machine, by which the work on four
barrels could be done at one time, and thus the results
be quadrupled.   His improvement in the slide lathe,
patented in 1855, is now almost universally adopted wherever this machine is used.   Among his patents assigned
to the Colt Patent Fire-Arms Manufacturing Company, are
three for making and one for packing cartridges; and
among his unpatented machines, is one for shaping the
barrel, and another for shaping the stock of the pistol.
In the construction of the buildings of the Armory, his
inventive genius devised methods which effected great
saving of timber and labor.   For the pumping apparatus
used in supplying water to the reservoir of Armsmear,
as well as in that used by the city of Hartford, Mr. Root
made several important improvements, which were intro-

duced into the steam-engines manufactured by Col. Colt for the Russian government.

Wherever and by whomsoever employed, Mr. Root was not content to do the work in hand as well as any other person under the same circumstances, but his thoughtful and ingenious mind was busy in devising a safer, shorter, more economical, and more efficient way of accomplishing the same or a better result. Possessing not only the gift of invention, as rare in mechanics as in poetry, which defies analysis, he had that other quality, equally rare, of consummate prudence or wisdom which discerned the line, invisible even to many brilliant minds, that divides the possible from the visionary. Hence every one of his patents had an immediate practical value, and his many suggestions on any subject in the field of his study and experience, even those which were not patented, were always prized. The two great companies whose operations have proved so profitable to their proprietors and stockholders—the whole community, that is benefited and enriched by the possession of better and cheaper implements in consequence of his witty devices—owe him not only high estimation and pecuniary reward while living, but, in memory, a large debt of gratitude, as one of the world's benefactors. To be appreciated as he was by the founders of these companies —to rise in their employ from the anvil and the lathe to the largest salary paid in the State, and to accumulate a handsome fortune for his family—was equally creditable to him and to them.

Mr. Root was eminently a man of modest stillness and humility—of many thoughts and few words. They tell a tale, says Lord Bacon, of a Spanish ambassador that was brought to see the treasury of St. Mark, in Venice, and still he looked down to the ground, and, being asked why he so looked down, said "he was looking to see whether their treasure had any *root*, so that, if it were spent, it would grow again, as his master's had in the mines of America." Well was it for the Patent Fire-Arms Company, that when death had snatched away its founder from his unfinished enterprise, and fire had done its worst, they had such a *Root*, in more than one sense. Well is it for them to-day that, though these inventors are dead, their ideas live incarnate in steel; and that, when their present incarnations shall have perished, their ideas will have but begun their posthumous life, in states unborn and accents yet unknown.

POTSDAM.

THE WILLOW FACTORY.

" Here thou wouldst think some fairy's hand
'Twixt osiers straight the willow wand,
In many a freakish knot had twined,
And foliaged tracery combined."

WHILE imitating the earthen dykes of Holland, Col. Colt was naturally led to borrow also their protective willows, and thus became the first planter on a grand scale of European willows in America. His first thought was merely to preserve his embankment from disintegrating through the dash and permeation of overflowing waters. The species of willow introduced for this purpose was the gray, or brindled, *Salix viminalis*, sometimes termed the French osier, as in this instance really brought from France, although equally common beyond the Rhine. Slips taken from the Colt plantation are to-day flourishing not only by many Eastern water-courses, but also in California.

. No sooner had the dyke become well fringed with the green and gray foliage, than a stranger appeared and wished to buy at a high price the willow crop. He had been in the habit of importing willows from abroad for manufacture, and it had occurred to him that he might

34

as well purchase them at home.   This hint was not
lost. It resulted in the establishment of the Colt Willow-
ware Works, which for several years have furnished employment to about six-score of operatives, some of them
children of armorers. The occupation is one more neat,
healthful, and educative than most sedentary pursuits. It
adds its quota to the resourcefulness of the community,
which is always in proportion to the varieties of its
industry. The more such varieties are multiplied, the
more sure is every man to secure the employment for
which he has special aptitudes, and hence to escape wast-
ing his vigor through mistaking his calling.

No characteristic of Col. Colt was more prominent
than that the firstlings of his heart were still the firstlings of his hand, and that he did nothing by halves.
Accordingly, no sooner had he resolved to utilize his
willows, than he secured skilled laborers, that he might
give his tools to those who knew how to use them.
With this end in view, he imported a whole village of
willow-workers from the fatherland of the willows. In
the southeast corner of his reclaimed land, and in immediate proximity to a dyke like those where they were
born, he reared them houses in their native style, a sort
of architecture sometimes called Swiss, but which is not
uncommon in other Teutonic regions, particularly in a
hamlet erected by a king of Prussia for a band of imported Russians. This hamlet is at Potsdam, and hence,
probably, the name Potsdam was given to Colt's willow-
world. The style of building at Potsdam is said to have
been adopted to please a Prussian queen, who, on her
way home from the Alps, had wished for a Swiss cottage, and on arriving at Sans Souci was greeted by the

sight of a whole village of such houses, which at the command of her husband had risen like a section of Switzerland spirited away to the far north. The continuation of Wawarme Avenue, on which it lies, might be significantly called by the name of the most famous of Roman roads in Great Britain, Watling Street. This ancient highway took its name perhaps from that of a man —possibly its builder; but in the speech of the people the word was soon corrupted into one unconsciously allusive to a natural effect of roads,—an effect more manifest the better they become,—namely, to weave or wattle far cities and far countries, year by year, into a closer and closer web of union.

In all countries the first roads, if called Watling-streets, would be descriptive names, inasmuch as no savages are found unskilled in wattling-works. Indeed, osier-weaving not improbably preceded in time that of wool or of hair, as much as picture-writing was before the alphabet.

The founder of the Willow colony further catered to German tastes by providing a beer and coffee garden, as well as by generous aid to the musical band which still bears his name, but which was started, before this colony was projected, to minister to the enjoyment of the whole community.

The most spacious of these tenements in the American Potsdam is large enough to accommodate or discommode eighteen families—a swarming bee-hive; a dozen others are each intended for only two. Their material is brick laid in a frame of wood: walls of this sort are cheaper, as they need not be of more than half the usual thickness. To an American eye these dwellings seem turned wrongside out, for the staircases are all outside, and the eaves overhang very far. In our Southern States, where chimneys are reared outside of the walls of houses, one reason assigned is, that there is more room there.

The Willow-ware Factory is not only worth seeing but going to see. The heavy work is of course done by machinery—the drudging slave, and soon, let us hope, the only one, of the nineteenth century. The willow wands, after steaming awhile, are thrust between two rollers, and there squeezed so hard that their bark becomes as loose as the skin of that fox which Baron Munchausen tied up and flogged, till the artful dodger escaped his tormentor by actually leaping out of his own hide through the hole of his mouth. Having been peeled, the rods are next split of all needed sizes in another machine, and sometimes planed in a third. They are then sorted, bundled, and laid in store for use.

The subsequent processes are greatly subdivided, in conformity to that principle of political economy which has been illustrated in our account of the Armory. For rough and heavy work the Irish operatives bear the palm. This sort of dexterity may be a Celtic heir-loom, for Cæsar tells us that the Celts of his time wattled Druidic images of wicker-work so huge and tough that men were incarcerated in them, and there burned alive in sacrifice. On the other hand, in all light and tasteful departments German workmen have the preference. All have patterns according to which they frame their handiwork, and a sort of lasts on which they weave it thread by thread. Sometimes a score of hands have to do with a single piece as it goes on from its first embryo to its last perfection. One man merely saws the notches in the frame-work of a chair, which by the way is made out of American osier, as stiffer than the kinds naturalized from abroad. Another tacks on the rim—a sort of wagon-tire—of East Indian ratan. He does not drive in the nails with a hammer, but squeezes them in by a pair of rough-jawed pliers, and that with a nimbleness amazing to an outsider. Other manipulations are executed after a fashion which every novice would pronounce foolish, if he did not see how well they turn out. In scrutinizing such handicrafts, we learn that there is more reason than we can at first perceive in other mysteries as well as in the roasting of eggs.

In many sorts of willow-work there is a four-sided oblong piece inlaid—alike to strengthen and embellish. For making this piece it is one man's business to bend sticks which have been soaked in water over night, force them into a frame which defines their size, and nail them in that shape. At first you feel sure that he is a bungler, and breaks each corner short off, but on trial you find it as strong as ever. Such swift-handed adroitness is ministered by practice, that he completes a score of the parallelograms in an hour. Again, some of the fabrics are decorated with a tapering volute of braid, which you think must cost a great deal of labor, but you will learn that a hundred yards of it, "small by degrees and beautifully less," are deemed no hard day's work. The differences in dexterity that are requisite for the different varieties of workmanship are great. A raw boy can soon manage to do the "siding up"—putting in as it were the woof—of a reticule; while the interlacing on the edge of its cover is so intricate as to demand several years of apprenticeship, and almost deserves for descri-

bing it Johnson's magniloquent definition of net-work, as
"any thing reticulated or decussated, with interstices at
equal distances at the intersections."

The more costly articles, having been fully elaborated,
go to the bleachery and there grow doubly fair. Those
goods which, proving unsalable, become shop-worn and
sallow, are sometimes bleached again, but more generally,
in virtue of dyeing and varnishing, come forth once more
as good as new. Who will invent us a machine from
which maids, yes, or men, now in the afternoon of their
best days, shall be turned out with wrinkles and hoari-
ness sloughed off, and yielding their gloomy sway to
gladsome rejuvenescence?

The wares here made grow in variety every year.
Those now advertised are of half a hundred sorts, and
strike a stranger who walks through the magazines as
even more numerous. Some of those least looked for are
chandeliers, or, as men are now beginning to say, gase-
liers, fenders, sleigh-bodies, picture-frames, chains for suspension baskets, stretchers, arbors, sofas.   Invalid chairs
are among the most luxurious specimens. The go-carts
make old bachelors resolve to have them the next time
they must learn to walk.

Chairs so easy to handle, looking too weak and ghost-
like to stand alone, as well as reminding us of Queen
Mab's turnout, where the wagon spokes were made of
long spinner's legs, yet safely sustaining men as un-
bounded in stomach as Falstaff, cannot fail to win their
way. The real strength coupled with apparent fragility
in these fabrics is so wonderful, that we recur to the
whole passage in the Lay of the Last Minstrel, from
which our motto was taken, for a fit expression of our

surprise, altering it a little, since no comparison goes on all fours :—

> " Thou wouldst have thought some fairy's hand
> 'Twixt osiers straight the willow wand,
> In many a freakish knot had twined,
> Then weaved a charm when the work was done,
> And changed the willow wreaths to stone,
> To slender shafts of shapely stone
> By foliaged tracery combined."

Willow furniture is cheaper than any other which is equally tasteful and serviceable. It can be transported to and fro through magnificent distances with less expense and injury, and has hence growing attractions for a race so migratory as ours, and above all for the millions who are nomadic in the boundless West. It is no wonder then that more than one willow factory has been started even in Wisconsin, while the Hartford Company, thanks to its heavier capital, its more perfect division of labor, and its more diversified patterns, can still compete with Westerners in their own markets.

Furniture of such a make that it yields on pressure, and so is soft as a bed of down, while it is cool and self-ventilated at the same time, has an indescribable charm for dwellers in the tropics, and to the rest of men at such seasons as the tropics visit them. Hence, a good deal of the Hartford fabric finds its way to Cuba, and the purpose of Col. Colt was to push the sale of it throughout all South America. Nor can it be doubted that through proffering Sleepy Hollow chairs and other household appurtenances, happily adapted alike to the torrid climate and the hardship-hating tastes of the Creoles, he would have secured rich returns.

In few things is modern improvement more manifest

than in the contrast between chairs of wattled willow,
which are sore labor's bath, as well as so luxurious that
they render ease more easy, and those portrayed by
Cowper :—

> "When restless was the chair; the back erect
> Distressed the weary loins that felt no ease;
> The elbows, rude and not with easy slope
> Receding wide, then pressed against the ribs,
> And bruised the side, and, elevated high,
> Taught the raised shoulders to invade the ears;
> The slippery feet betrayed the sliding part
> That pressed it, and the feet hung dangling down,
> Anxious in vain to find the distant floor."

TORPEDOES.

TORPEDOES.

"His unda dehiscens,
Terram inter fluctus aperit, præruptus aquæ mons."

"Leviathan causeth the deep to boil like a pot; the fountains of the
great deep are broken up."

MILTON thought his Paradise Regained a superior
poem to his Paradise Lost, and most parents betray a
lurking fondness for their most unlucky child. On a
similar principle, Col. Colt's pet idea was not his
revolver, which all the world lauded, but rather his
submarine battery, or torpedo, which all the world
laughed at. The difference in the two cases is this:
that while the world still· prefers Paradise Lost to the
other poem, it is already beginning to admit that torpedoes are of more tremendous efficiency than revolvers.

When the late inventive and indefatigable but most
unassuming President of the Repeating Fire-Arms Company, E. K. Root, Esq., among the earliest as well as
latest friends of Col. Colt, was asked about his first inter-
view with that gentleman, "It was," said he, "at Ware
Pond, on a Fourth of July, I think in 1829. It had
been noised around that a youngster—one Sam. Colt—

would blow up a raft on the pond that day, and so I
with other apprentices of the neighborhood walked some
way to see the sight.   An explosion was produced, but
the raft was by no means blown sky-high.   Yet, curious
regarding the boy's explosive contrivances, I then and
there made his acquaintance."

In the light of future developments, it is easy for us
now to see that these holiday freaks with a little powder
on a country pond were nothing less than

> The baby-figure of a giant mass,
> Henceforth to come at large.

In his letter to President Tyler, Col. Colt describes himself all along, from the period of this first meeting with
Mr. Root, as "employing his time in study and experiment
to perfect" his submarine explosives.   For some time,
however, both before and after the year 1836, when he
was granted the earliest patent for the revolver, his
labors and anxieties concerning that weapon and its
fabrication cannot have left him much of either time
or strength for any other pursuit.   But in 1841, when
his Paterson company had failed, and he could find no
other capitalists who would risk a single dollar to help
him realize his idea of a million-minting revolver, he
returned with new zest to aquatic pyrotechnics, as it
were to a first love.

All his previous earnings had been swallowed up in
abortive arms-factories; he therefore solicited govern-
mental aid for trying submarine experiments which were
beyond his private means, and which, if successful, were
sure of redounding to the public good.   With this view,
repairing to Washington, he addressed the following
letter to the Chief Magistrate of the country :—

WASHINGTON, *June* 19, 1841.

SIR:—It is with no little diffidence that I venture to submit the following for your consideration, feeling as I do that its apparent extravagance may prevent you from paying it that attention which it merits; and but for the duty I owe my country in these threatening times, I should still longer delay making this communication.

For more than ten years past I have employed my leisure in study and experiment to perfect the invention of which I now consider myself master; and which, if adopted for the service of our Government, will not only save them millions of outlay for the construction of means of defence, but in the event of foreign war, it will prove a perfect safeguard against all the combined fleets of Europe, without exposing the lives of our citizens.

There seems to prevail at this time with all parties a sense of the importance of effectually protecting our sea-coast; and as economy is a primary consideration in the present exhausted state of our treasury, I think I have a right to expect a favorable consideration of the propositions which I have determined to make.

By referring to the Navy State Papers, page 211, you will discover that Robert Fulton made experiments which proved that a certain quantity of gunpowder discharged under the bottom of a ship would produce her instant destruction. That discovery laid the foundation for my present plan of harbor defence; and notwithstanding the failure of Fulton to use his invention to much advantage, in its imperfect condition, during the last war, one glance at what he did perform is sufficient to convince the most incredulous, that if his engine could be brought under easy and safe control, it must prove an irresistible barrier against foreign invasion.

Discoveries since Fulton's time, combined with an invention original with myself, enable me to effect the instant destruction of either ships or steamers at my pleasure, on their entering a harbor, whether singly or in whole fleets; while those vessels to which I am disposed to allow a passage are secure from a possibility of being injured. All this I can do while myself in perfect security, and without giving an invading enemy the slightest sign of his danger.

The whole expense of protecting a harbor like that of New York would be less than the cost of a single steamship; and when the apparatus is once prepared, one single man is sufficient to manage the destroying agent against any fleet that Europe can send.

With the above statements as an intimation of what can be done, I will mention in as brief a manner as possible the terms on which I will make an exhibition, to prove to yourself and your Cabinet that a sailing vessel or steamboat cannot pass (without permission) either in or out of a harbor where my engines of destruction are employed.

To make the exhibition (which I contemplate, should I meet with sufficient encouragement) will require an expenditure of $20,000, which sum I will employ for that purpose from my own means, on condition that the Government will lend me such aid as I shall require (which can be supplied without incurring new expenses), and that when I get through my exhibition the Government shall refund to me all money which I shall have expended, and pay me an annual sum as a premium for my secret.

I hope I may be excused from mentioning that, as any hint of my plans at this time must prove prejudicial, it is my wish that the present communication may be kept from the view of all persons excepting the members of your Cabinet. I have the honor to be, Sir, most respectfully,

Your Excellency's devoted

And obedient servant,

Samuel Colt.

It does not appear that this epistle elicited any immediate favors either from Mr. Tyler or from any member of his Cabinet, yet, like the importunate widow, he persisted in his prayers till he gained his first object. Hence, in the *New York Herald* for March 17, 1842, we find it stated that Mr. Colt, inventor of the revolving pistol, had been already, for some time, engaged, under the authority of the Secretary of War, in experimenting, and that it was asserted he could ignite a destructive shell under the water, at the distance of ten miles, in a few seconds, its principle being founded on the electric fluid.

On the fourth of June that same year,—as contemporary newspapers record,—having sunk a case of powder in New York harbor, he threw up therewith as it were a waterspout to an amazing height, and that while himself standing at a great distance.

On the fourth day of the next month, July, an explosion produced by Colt's newly invented submarine battery,

in the bay opposite Castle Garden, was chronicled in
all the papers. One of Colt's torpedoes threw up out
of water an old gunboat, the Boxer, and split its timbers to kindling-wood. The inventor of this fire-monster,
having moored it more than a dozen feet beneath the
bottom of the vessel, touched it off, while himself on
the deck of a United States war-ship at some distance.

The popular impressions produced by this beginning of
experiments are apparent from the following contemporary
letter, which may be still seen in the Colt archives:—

LETTER FROM JOHN MOUNT TO COL. COLT.

JERSEY CITY, *July* 6, 1842.

MY DEAR SIR:—I cannot refrain from writing you
my congratulations on the entire success of your recent
submarine explosion. I witnessed it, among hundreds,
from the lower wharf of Jersey City. As the dense
volume rose heavenward, its terrific grandeur could only
be exceeded by the amazement and wonder of the multitude around me at the means by which it was accomplished. Its altitude, in a line of vision from where
I stood, was far above the royal truck of the British
frigate Warspite, consequently must have been very high.
I trust, my dear sir, that the Government will properly
appreciate the vast importance of this mode of defence,
and that you may reap the honors and emoluments to
which you are justly entitled.

Thine in sincerity,

JOHN MOUNT.

So decided was the success of the explosive effort on

Independence Day, that a schooner on the Potomac was immediately offered by the Government for Col. Colt to destroy, " if he could." He accepted the challenge, and was confessed on his first trial to have thrown it sky-high, on the twentieth of the month following, namely, August, 1842. This exploit he performed in the presence of the President, Heads of Departments, and Gen. Scott, and that while himself stationed at no less than five miles from the vessel he demolished. As soon as the ship's destruction was complete and indubitable, the steamer bearing the Presidential party was headed to the station where Col. Colt had pulled his electric trigger; the operator was invited on board, congratulated by the distinguished personages, and greeted with a bouquet from the hand of the President's young daughter, Miss Tyler. This floral tribute, its roses, pansies, cedar, and amaranths, still well preserved, holds the place of honor to this day in a scrap-book which Col. Colt filled with odds and ends from all quarters relating to his torpedo experience.

It is no wonder that all souvenirs of this day were precious in the eyes of the experimenter. He had run a gauntlet of many revolver failures through many a year. As a projector of submarine explosives also, his way had been through a hedge of thorns. But now, for the first time in so many years, smooth success seemed strewed before his feet. Well might he be exultant.

His first Potomac experiment took place on the twentieth of August, and before the end of that month a joint resolution had passed both houses of Congress for granting $15,000 in furtherance of more thorough submarine tests and trials. In the words of a contemporary letter,

36

"The explosion carried the house by storm, notwithstanding the captious opposition of Mr. Adams."

Ten days before his Potomac triumph, Colt had written to Mr. Adams in the following words:—

GADSBY'S HOTEL, *August* 11, 1842.

SIR:—I am anxious for an opportunity to converse with you on the subject of my plans for harbor and coast defence by means of a submarine battery. If you can make it convenient to appropriate a small portion of your valuable time to a subject which I think bids fair to result in important changes in the defence of our maritime frontier, I shall be exceedingly happy to wait upon you at any time and place you shall direct. With great respect,

I have the honor to be, Sir,

Your most obedient servant,

SAMUEL COLT.

Appended to the copy of this note, still preserved in one of the scrap-books, is the laconic remark:—

"N. B. To the above letter no answer was received.

"S. COLT."

Mr. Adams in the House decried rendering aid in testing subaqueous batteries. He also used his influence in the Senate against such a measure. His antipathy to projects of such a nature was ascribed by some to his being seventy-five years old, since all English doctors, who were over fifty years of age when Harvey discovered the circulation of the blood, died denying its motion.

Others imputed his disfavor to the noise of one of Colt's experiments having frightened Mr. Adams's horses, so that they ran and injured the coachman who was driving his empty carriage.

At all events, the arguments employed against the appropriation for experiments were unworthy of their illustrious author. One, that "Colt could not accomplish any thing," was disproved on a fresh trial; so that Mr. Adams's ridicule of "throwing money into the sea" turned out no less absurd than it would be to stigmatize the fisherman's baiting his hook as throwing food into the water.

His other objection, namely, that even if he could use torpedoes destructively he would not, because it would be "cowardly, and no fair or honest warfare," is identical in spirit with that already commented on in this memoir as urged by Don Quixote against all use whatever of villanous saltpetre. It would have been annihilated in a torpedo-shock of laughter by a simple reference to the tirade of the Knight of La Mancha. It is quite in the vein of the bald, unjointed chat already quoted from the popinjay who said:—

> "It was in sooth great pity, so it was,
> This villanous saltpetre should be digged
> Out of the bowels of the harmless earth,
> Which many a good tall fellow had destroyed
> So cowardly, and but for these vile guns
> He would himself have been a soldier."

Mines under the earth have been always and everywhere deemed justifiable. What valid objection can there be to mines under the water?

So little were the objections of Mr. Adams regarded, that Col. Colt saw the appropriation for a thorough scrutiny of his pyrotechnic pretensions passed by a majority sufficient, had there been need, to override a veto. Elated by this auspicious sign, he at once girded himself for showing still more signal proofs of his skill.

Accordingly, on the eighteenth of October, 1842, the Volta, a brig of three hundred tons, was blown up in New York by his infernal machines. This experiment was conducted under the patronage of the American Institute, in connection with their annual exhibition, and before spectators who were estimated in the *Tribune* to number forty thousand. At a given signal the hulk, through some unseen power, seemed lifted from the water by its waist, the bow and stern sunk heavily, and the whole was enveloped by a huge pile of dense mist. In about a minute it all settled down, and a large circle was strewed thickly over with fragments of the doomed vessel, the largest piece being a slight portion of the hull attached to the mainmast. Among the witnesses of this scene, on the ship of the line North Carolina, was the Secretary of War. On this occasion Mr. Colt applied the electric spark while at some distance, and on the revenue cutter Ewing.

Not satisfied with destroying vessels when at anchor, and that while himself beyond the range of their guns, the inventor of the submarine battery would in the next place demonstrate that it was no less able to sink vessels while under sail.

His grand trial of this feat was made in the Potomac, near the Arsenal, on the thirteenth of April, 1844. Congress adjourned to witness it, and they saw a ship of five hundred tons, while sailing at least five knots an hour, almost immediately after the firing of a rocket as a signal, ingulfed as if through the force of a water-earthquake, and that by the hand of an operator who was five miles away. In the judgment of the naval officer in charge of the vessel thus devoured, the British squadron would have shared its fate in 1814, and thus our Capital have been saved from capture, had the Potomac been at that crisis defended by Colt's contrivances. The shipwreck thus accomplished was the more unaccountable, because it could not have been known beforehand to Col. Colt, even on the supposition of collusion, which way the veering of the wind would make it necessary for the vessel to sail, and because that vessel had been deserted by all on board some little time before it was shivered to atoms. Probably torpedoes had been planted in the channel both above and below, and, it may be, all around it, and that possibly in more than one circle.

The good is always the enemy of the better, and so naval officers had one and all predicted failure. Commodore Perry said, that if the congressional appropriation were to be divided among those who would volunteer to navigate a ship safely through Colt's gins and pitfalls, volunteers would not be wanting for that endeavor, while if Colt could blow them up he would deserve the money, and they would not need it; but that he was sure to fail. Therefore, onward from his glorified hour on the Potomac, the experimenter himself remained all the more satisfied that his was not merely the spiriting of the tricksy Ariel, who flamed amazement in waist, and deck, and cabin, but the rough magic of Prospero, who

" 'Twixt the green sea and the azure vault
  Set roaring war, and rifted Jove's stout oak
  With his own bolt, "

till mariners cried, "Hell is empty, and all its devils
are here!"

The principle of his achievement, as explained at the
time by prying reporters, is simply as follows:—A float-
ing cylinder filled with powder is moored in a channel
where a ship must pass near or over it.   Running
through the powder are two wires, united in the midst
of it by a filament of platinum, isolated by means of
silk bound round them, and then coated with a resinous
varnish.   The ends of these wires communicate with a
voltaic battery on shore.   At the instant we notice a
vessel to be directly over the cylindrical magazine, if
we complete the circle of the electric fluid, as it darts
through the powder it ignites it, and the ship is
destroyed.

This, however, is a shallow and unappreciative view of
Col. Colt's submarine doings.   There is good reason to
judge him the first man who devised and laid down an
insulated marine telegraphic cable.   On this point the
Hon. Henry C. Deming thus spoke before the Agricultural Society of Connecticut, in 1856: "In the winter
of the year 1842–3, a citizen of Hartford, the same who,
without loan or discount from the banks, carries on his
single shoulders the tremendous burden of our South
Meadow improvement,—but at the time of which I speak
so poor that he could scarcely call his mathematical
instruments, or even his watch, his own,—he, sir, laid
down on the bottom of the East River, near the line of
the Fulton Ferry, a submarine cable, and, higher, another

at Hell Gate, which differed only from the one recently lost between Nova Scotia and Newfoundland in this respect, that in the latter gutta percha was used as an insulator, whereas in the former (gutta percha being then unknown) cotton yarn, with a combination of asphaltum and beeswax, wound about the wires, inclosed in a metallic tube, was used as an insulator. He actually had in working order, between Coney Island and the Merchants' Exchange, an electric telegraph in which this submarine cable was a part of the communication. I therefore file my caveat here in behalf of Col. Samuel Colt, of the city of Hartford, for the invention of the submarine cable." Again, according to Col. Totten, an unwilling witness—"No project, so far as he knew or believed, had ever, antecedent to Mr. Colt's, been tried in practice for communicating fire by galvanic electricity to deposits of powder placed in harbors for their defence." So, too, Prof. Henry says, that though Prof. Hare of Philadelphia was first to suggest, and perhaps illustrate in his classroom, the application of galvanism for producing subaqueous explosions, yet "Mr. Colt deserved great credit for the industry and practical skill with which he brought the matter before the public."

In addition to this, Col. Colt claimed to possess a further secret, which many of his friends think died with him. But so reticent was he on the subject, that to this day it is not clear whether this secret related to the explosive compounds, or to the mode of arranging them, or to ascertaining at what instant it was necessary to fire the aquatic mines. One of his projects may have been to compel a vessel which touched his explosives to tell, and that by telegraph, its proximity.

"Old Neptune keeps his secrets well," and so it seems do those who invade his fastnesses.

However marvellous the ship-shattering shocks, they failed to secure the introduction of torpedoes as an arm of national defence; and although the committee on naval affairs reported a bill for his encouragement on the eleventh of January, 1845, and although, on the first day of that month, while in Baltimore, he had exploded powder forty miles off, in Washington, yet the projector appears about that date to have diverted his attention from submarine engines, and to have turned his energies anew to the revolver.

After all, to the end of his life, he was now and then fondly brooding over "the first heir of his invention," water-mines.

He had this fondness in common with Robert Fulton. He loved to follow Fulton, in 1797, trying submarine shells on the Seine under the patronage of Napoleon; in 1804, entering into articles of agreement with Pitt to destroy the French navy; in 1805, making assaults on the Boulogne flotilla; in 1810, granted five thousand dollars by the Congress of the United States, and demolishing a ship at New York, yet, in spite of all, failing to induce the Government to adopt his new-fangled weapons.

Indeed, it must be confessed that, up to the time of our Great Rebellion, torpedoes had not been proved of much practical value.

The name torpedo, as denoting a subaqueous explosive, was unknown to Johnson when he made his last revision, in 1778. We probably owe the word to Fulton, whose chief pride was to have invented the thing. He seems to have so named it because it stuns, or makes *torpid,*

and that a long way through water. In so naming it
"he builded better than he knew," for Colt's variety,
being fired by electricity, may with special fitness take
its name from the electrical eel.

One subaqueous contrivance in the Revolutionary War,
called Bushnell's Marine Turtle, alarmed, but failed to
injure the British squadron in New York.   Another,
employed against them in Philadelphia, missed indeed
its particular mark, but did not fail to revive fainting
patriotism, for it inspired Francis Hopkinson with his
ludicrous ditty on "the Battle of the Kegs," an effusion
which, sung to the tune of Moggy Lawder, was more
mirth-moving in the days of '76 than any one comic song
of recent years.   During the war of 1812, several trials of
torpedoes in actual attacks failed of their object. But
this record did not discourage Colt. His motto was,—

> "At last the Greeks got into Troy. There's no denying,
> All things are done, as they did this, by trying."

He saw progress. He supplied some links which were
lacking in the chain cable on which others had hoped
to outride adverse blasts.   He applied in a new way
the telegraphic wire which, at this present writing, has
just made the two hemispheres Siamese twins,—but
which he, now nearly twenty-four years ago, was the
first of men to manufacture, and to insulate under water.
Honor to him for his high ideal of mechanical and
chemical capabilities! For now more than a decade,
facts have shown a progressive tendency to confirm
his theory of harbor defence.   Nor is it longer to be
disputed that, if not in heaven and earth, at least in the
waters under the earth, there are more things possible

37

than were dreamed of in the philosophy of the routine
conservatives, who denied the submarine visionary room
and verge enough for making full proof of a defensive
resource always potent, and sometimes omnipotent. During the Crimean war, torpedoes were anchored in the
channels leading to Cronstadt. Such was their construction, that a ship which struck one of them would touch
it off, and thus commit suicide. Who can say how
much influence these threatening engines had in deterring
both French and English from approaching the key to
the Russian capital? Yet many still held that, like the
Quaker cannon with which we are of late familiar, however mighty to scare, powder sunk in the water was
impotent to harm. But now, at last, no man dares longer
deride subaqueous destructives. Too provokingly, in
many a Southern port, have they defied for years our
blockaders. Such implements of destruction, in the course
of the late rebellion, sunk the Harvest Moon, Admiral
Dahlgren's flag-ship, on the Pedee; they sunk three
transports on the St. John's; they sunk two of our mon-
itors and four wooden gunboats at the entrance of
Blakely River, although the river had been thoroughly
dragged, and many torpedoes were removed before the
ships went over the bar. They were the main reliance
of the rebels for the defence of Charleston, and they prevented our navy from running its batteries. On the
15th of January, 1865, the iron-clad Patapsco, venturing
forward on a reconnoissance, was so disintegrated by a
torpedo that, within half a minute after the explosion,
her deck was under water, and two-thirds of the crew
went down with her. The Bibb, the Massachusetts,
and the New Ironsides, though not fairly hit, were shat-

tered or shocked near the same harbor. Similar statements in reference to the efficiency of subaqueous mines
are scattered through the reports of our naval operations
in Southern waters. Even at this writing, August, 1866,
tidings comes that a forgotten rebel torpedo at Charleston
has just exploded, destroying a ship, and not without loss
of life.

In view of such facts, Admiral Dahlgren declares in
his official report that, " had our fleet sailed into Charleston harbor, the submarine explosives would have proved
as troublesome as it was expected they would be," that
is, by those who feared them too much to approach
them; the Secretary of the Navy adds, that torpedoes are
always formidable, and that, during the late rebellion,
they have been more destructive to our vessels than all
other means combined ; Commodore Porter's answer to
inquiries from the head of the engineer corps is, that
when combined with obstructions, they are a better
defence than any of our present forts; and the last *North
American Review* closes a thorough article on the mechanics of naval warfare by saying :—" We believe torpedoes are to play a momentous part in the warfare of the
future. Their power has hardly begun to be developed."
Col. Colt would have counted it worth half his life to
behold these exploits of submarine batteries, forming so
striking a fulfilment of his prophecies, and to read
these opinions of the highest authorities contradicting
those by means of which, twenty years before, Col. Totten and the then Secretary of War cut off from him
that Government patronage which would have realized
his submarine ideal no less triumphantly than his own
unassisted energy realized that of the revolver.

The torpedo theory, as it was when Colt felt constrained to leave it, reminds us of Milton's lion, who is
described as emerging from the earth still only half
created, and pawing to get free his hinder parts. Now,
that same theory resembles that same lion when he

> "Springs, as broke from bonds,
> And rampant shakes his brinded mane."

MEMOIR.

MEMOIR.

Samuel Colt was born at Hartford, July 19th, 1814. He was the third son of Christopher and Sarah Colt. His mother was a daughter of Major John Caldwell, one of the most prominent citizens in the city of Hartford of that day.

Samuel inherited from his mother and maternal grandfather his most marked and characteristic traits of mind; and, if we may judge from the miniature of the former, from her too he inherited his regular, beautiful features. Before he had completed his seventh year, it pleased God to take from earth the mother he so tenderly loved—and her place in this son's heart was always filled with the most gentle, loving memories. Only a few weeks before his own death he was having some copies of his mother's picture prepared, one of which was for his room, where were gathered the pictured forms and memorials of those he loved best, and among these the treasured mementoes of the mother and sisters, who long years ago had passed from this life. It was but a little while after that dear mother died, that the work of life for this young boy began. Before her death, his father, who had once been among the wealthy citizens of Hartford, became embar-

rassed in his business affairs, and eventually lost the
bulk of his property. I have heard the son relating
how, when the news of this misfortune was told his
mother, he, a little child, was in the room playing
under the piano. With clasped hands, and eyes dimmed
with tears, she only said, "My poor little children!"
Those simple words were uttered more than forty years
ago, and yet for all that time they showed to the boy,
and to the man, how the mother had loved her children
more than herself, and feared poverty most for her
darlings.

She died of consumption. On a fair June day, the
little lad went out on the hill-side to gather strawberries
for the invalid mother. While his willing hands were
working diligently, in obedience to the promptings of his
loving heart, the mother's spirit passed away, beyond
the reach of earthly loves, as well as earthly wants and
cares. But who shall say that that mother's spirit was
not ever near the son, who never forgot her, to be his
guard and his shield?

For a short time after the mother's death, the children
were under the care of Mrs. Price, a sister of Mr. Colt.
Dear Aunt Price! her kind, motherly eyes soon saw all
the merit and the strong character that was lurking behind the fun and roguish tricks of the boy, so warm-
hearted and impulsive. They were ever firm friends,
and many a time she stood between him and punish-
ment. He always spoke of her with respect and gratitude, and, shortly before her death, addressed to her an
affectionate letter, in which he told her that he hoped
very soon to be able to gratify one of his earliest and
favorite ambitions, and place the dear aunt whose kind-

ness had befriended his boyhood, in a pleasant home of
her own.

There was another sister of his father's, Mrs. J. D. Selden, of Troy, a lady of an uncommonly fine intellect and
great vigor of mind, who watched the development of his
character and genius with intense interest. She, I believe,
rendered him some pecuniary assistance, as a loan, when
buying back his patent-rights, which he had disposed of
to the Paterson Company. And as he rose step by step
in fame and fortune, her heart bounded with pride and
pleasure that one of her dear brother's sons was en-
nobling the name she bore in her youth. Though past
the age of eighty-five, she came for several successive
years to visit him at Armsmear, where at each visit she
saw some new beauty developed. But he went down to
the grave before her, she living until she was nearly
ninety-three, with her mental faculties but slightly impaired, her interest in public affairs unabated, and her
heart still warm and true to those she loved. Early in
the summer of 1864 she rested from her labors.

Mrs. Colt died Saturday, June 16, 1821, when Samuel
had not quite reached the age of seven years. His father
was married a second time, to Miss Olive Sargeant, on
the 12th of March, 1823, before Samuel was nine years
old. At ten years of age, he was sent to his father's
factory at Ware, Massachusetts, where, with intervals
at school and on a farm, he remained until he was sent
to Amherst, for the purpose of extending his education.
With little taste for study, he yet learned rapidly all
practical branches of information within his reach, and
even in those days was a leader of all the boys, whether
in work or play.

38

Among the traditions of his boyhood, one is given by a neighbor on the hill, showing at how early an age his attention was directed toward the arm with which his name was to be so intimately connected, all the world over. When about seven years of age, he one day was for some time missing, and when at last discovered, he was sitting under a tree in the field, with a pistol taken entirely to pieces, the different parts carefully arranged around him, and which he was beginning to reconstruct. He soon, to his great delight, accomplished this feat. About two years later, he was sent to Glastenbury to work on a farm, for a year. From his own account, he did not find there a very gentle master, and certainly was in no danger of being spoiled by over indulgence. Yet doubtless the farm-life at that early age did much to develop the little frame, and laid the foundation of that rugged health he enjoyed for many years, before he was attacked by the pitiless gout.

A relative still remembers, in crossing the bridge over the Connecticut to East Hartford, being startled by the sharp report of a pistol ringing out on the clear morning air. Looking onward he discovered young Sam, stopping on the bridge to indulge in the delight of firing his pistol into the river, while on his way back to the farm, after a holiday at Hartford. These little incidents serve to prove how early in life his taste for fire-arms was shown; and how it "grew with his growth and strengthened with his strength," after years have fully demonstrated. In what flights of fancy he may even then have indulged we can only guess, and yet, bright as they may have been, were they not fully realized?

Both his sisters had died, one in 1825, the other in

1829. Other children were born to Mr. Colt, and Samuel's life became very early one of self-reliant exertion; but through all the hardships he underwent, amid struggles to overcome gigantic obstacles, the tender memories he ever cherished of the mother and sisters, whose love had blessed his earliest years, are to me one of the most touchingly beautiful traits in his strong character. When wealth and honors had crowned his struggles, and he had built for himself a home, there, in his own private room, the simple designs painted by his young sisters, neatly framed, hang upon its walls; and his mother's pictured face, with his maternal grandfather's, are side by side with those of his wife and his own little ones, so many of whom are among the angels.

The little room is solitary, now that the master has gone; there is the chair in which he last sat, the writing-table, the idle pens, and the silent bell. We turn to the window, and the saddened gaze falls upon the sweet spot he chose for the tomb of our first son, and where the father too now sleeps by his darlings.

His father had arranged for sending him to sea. Before the ship was ready to sail, he concluded that he would not wait to be sent for, and left school without leave from any one, after some patriotic demonstration on the Fourth of July which did not meet the approbation of the school authorities. Arriving at home unexpectedly, he told his father that he thought it must be near time for the ship to sail, and had come to see about it. "Have you brought all your things?" asked the father. "All but my bills," promptly answered young Sam.

He sailed from Boston on the second of August, 1830, in the ship Corlo, Captain Spalding, for Calcutta, as a sailor before the mast, and several missionaries went out on the same voyage. A lady member of the family still remembers being present when he made his farewell call upon his grandfather Caldwell, and his quizzical expression when he told the old gentleman that Aunt Price was perfectly easy about him now, since she knew the missionaries were going in the same ship.

Mr. Samuel Lawrence of Boston aided in selecting his outfit, and the following is an exact copy of the letter sent by Mr. Lawrence to Mr. Colt on the sailing of the ship:—

CASH PAID FOR S. COLT.

Seaman's cap	$3 50
Quadrant, almanac, and compass	18 50
Mattress, bedding, &c	9 00
Slop clothes	38 92
Boots and shoes	8 00
Stockings	2 00
Jackknife, &c	1 00
Custom House	25
Seaman's chest	4 62
	85 79
Cash	5 00
Paper, &c	45
	$91 24

Boston, *August* 2, 1830.

Dear Sir:—Above is a memorandum of sums paid for your son. It was necessary to be prepared for cold and warm weather, and in first fitting out there are a great many things necessary, which need not be replaced for years. Sam will not require any money of consequence; he may find something to bring home. I told the supercargo to advance him fifty dollars if he required

it. The ship sailed this morning. The last time I saw
Sam he was in tarpaulin, checked shirt, checked trousers, on the fore-topsail yard, loosing the topsail. This
was famous at first going off. He is a manly fellow, and
I have no doubt will do credit to all concerned. He
was in good spirits on departure. There were some
thousands present to see the missionaries off. Prayers
and singing were performed on board.

Your friend,

S. LAWRENCE.

He had a hard life on shipboard, and was most thoroughly cured of any wish to be a sailor by profession.
Yet he always loved the sea, and the frequent voyages
to Europe, which in after years his business compelled
him to make, were almost the only intervals of relaxa-
tion, lasting more than a few hours, that he found. On
this voyage to Calcutta, he first conceived the idea of
"Colt's Revolver," and commenced a little wooden model
on shipboard. The more he thought of his invention,
the more he became assured of its worth, and, without
money or friends to aid, he determined to go on and
make full proof of it.

Upon his return from sea, he was for a while again
at Ware, in his father's factory, in the dyeing and bleaching department, under Mr. William Smith, who possessed
much valuable and practical chemical knowledge. With
the information, obtained chiefly from Mr. Smith, for his
capital, he conceived the idea of lecturing on chemistry;
and in 1832 he commenced his lectures, administering
the laughing or nitrous oxide gas, and going into every
town of two thousand inhabitants in the United States,

Canada, and Nova Scotia, under the name of Dr. Coult.
The lectures seem to have been quite popular, judging
from the newspaper notices of them, and met with varying pecuniary success; but, besides paying his travelling
and other expenses, he managed to save enough to continue his experiments, and also to commence the fabrication of his arms in Baltimore, that is, one by one, as
he earned the money to pay for the work.   While in
Montreal and Quebec, the cholera broke out there very
alarmingly, thus materially lessening his profits.

Gradually working his way along, he went in 1835 to
Europe, for securing his patents there, and returned on
the seventh of January, 1836.   I have often heard him
say that, while in London during this visit, he lived on
only a shilling a day, and could afford no more.

He was at this time six feet in height, straight, and
very slender.   His face was uncommonly beautiful, with
very perfect features; clear, honest eyes of light hazel,
and with a wealth of the finest hair covering his head,
in short, crisp curls.   As he grew older, his figure developed into more massive proportions, seeming to keep
pace with the ever expanding, active brain.   As years
went on, and he began to feel the responsibilities of the
position to which he had raised himself,—when he found
his opinion sought for by the great, the wise, and the
good,—when even monarchs sought the benefit of his
wide experience and inventive genius, in their own national works,—when the endearing ties of home and
children had brought out into strong life all the gentleness and tenderness of his nature, there was painted on
that noble face a soul-beauty, that made it more charming than it was in all the glory of youthful days, even

though threads of silver were stealing in among the brown of the clustering locks.

In 1836 he commenced manufacturing arms at Paterson, New Jersey, with a company bearing the title of "Patent Arms Manufacturing Company," having a capital of three hundred thousand dollars.

His efforts at this period were unceasing for bringing the arms to the favorable notice of the Government. The Secretary of War, Mr. Poinsett, appears to have entertained a most excellent opinion of them, but the low state of the Treasury, and the failure of Congress to make sufficient appropriations for that department, deterred him from giving the much desired orders for adopting repeating arms into the service.

During the Indian war, in the winter of 1837 and 1838, Col. Colt went to Florida, carrying a few of his arms, which met with much favor from Gen. Jessup and Major Harney, then in command there, and requisitions were made for more. He passed a hard winter among the Floridian swamps and everglades, but made there the acquaintance of many officers, some of whom were life-long friends, and one of whom, Col. Charles May, came to pay friendship's last tribute, to look upon that noble face once more, and follow his body to the grave, which claimed that day so much that was great and true.

On the 10th of April, 1838, while going from Fort Jupiter to St. Augustine, the vessel was delayed by head winds, and Col. Colt, with two other gentlemen and a crew of four, started from her in a small boat for the beach. When about a mile from it, the boat swamped among the breakers. Thus he lost all his baggage, and was himself four hours in the water, until assistance came from the shore, narrowly escaping with his life.

He was also at the same time so unfortunate as to lose his pocket-book, containing, among other papers, a Government draft, belonging to the Company, which occasioned much serious inconvenience and blame, a long time elapsing before it could be replaced.

Having spent the winter of 1838 and 1839 in Washington, endeavoring to induce Government to give an order for his carbines and rifles, and failing to effect his object, he was obliged, for meeting his liabilities, to sell his patent to the Paterson Company. In May, 1840, a board of United States officers reported unanimously in favor of the arms. It was composed of Captains McCauley, Aulick, and Twiggs. But the Company became seriously embarrassed, and in 1842 failed, so that all manufacture of the Colt Arms was suspended.

The Submarine Battery had been already for some time receiving a share of Col. Colt's attention, and at length Government made an appropriation for the purpose of testing its worth; and in 1842, and following years, he made very successful experiments, totally destroying vessels when under full sail;—two of these took place off Castle Garden; the others at Washington, near the Arsenal, on the Potomac. Full and graphic accounts of these feats were published at the time. At one on the Potomac, President Tyler and family, and the Cabinet, were present, and the young daughter of the President there presented him with a bouquet, which, faded and brown, is still among his papers.

In the midst of these trials of explosives, he was engaged in the offing telegraph, and in 1842 and 1843 he constructed the submarine telegraph to Coney Island and Fire Island Light. This was the first submarine telegraph ever successfully laid and operated. He also laid a submarine telegraph from the Merchants' Exchange Reading Room to the mouth of the harbor, crossing several small streams of water, with telegraphic wires submerged and insulated. A portion of the cable used at New York, having been suspended on Broadway on the night of the great celebration for the laying of the Atlantic Cable, was stolen before morning. He used asphaltum and wax as insulators, the whole being inclosed in a leaden pipe. Afterwards, he received from London another portion of the same cable, which, some years before, he had presented to the Institute of Civil Engineers. As a pecuniary speculation, the offing telegraph, both in New York and Boston, failed.

In 1847 he commenced manufacturing his arms at Whitneyville, near New Haven, having at the outset of the Mexican war received an order from the Government, at the instance of Gen. Taylor, for one thousand pistols; and, although large numbers had been manufactured at Paterson, there was not at this time a single arm that could be procured in the market. The so-called Texan model, the Rangers soon made a terror to the Mexicans and all enemies, and of world-wide renown.

From this time forward fortune smiled, and success was his—all the more prized that it was purchased by industry, self-denial, perseverance,—by days and nights of weary toil and waiting. He bought back his patent-rights as soon as possible from the various parties owning them. Henceforth, with his affairs in his own control —with his experience, his sagacious, far-seeing mind,

39

he was sure to succeed, nobly, triumphantly. Yet, when his dreams were more than realized, he never forgot to be generous, and his house was ever open to his friends with genuine hospitality. The poor and the sick, the aged and infirm, he remembered out of his abundance. While his mind was clear as crystal and firm as granite, his heart was as affectionate and gentle as a woman's. I have never seen the two, strength and gentleness, so wonderfully combined in the same person. His tender care for the aged, and for little children,—his agony when called to part with his own darlings, I can never forget. His love of flowers amounted almost to a passion, and the green-houses were a source of untold happiness. Many a wintry morning, when the family gathered in the dining-room for breakfast, a lovely bunch of flowers would be laid at each plate, which he had gathered in his early walk, thus gracefully remembering all.

His care for his father's family, each one of whom he helped in many ways, even when he himself was poor, and struggling on toward the goal he had in view, and the princely munificence shown them in later years, have scarcely a parallel. In more than one sad instance, he heaped the most generous bounties on those who only repaid them with malice and slander of him and his. Yet he, the quick, high-tempered, impulsive man, said not a word in angry retort, and showed how deeply his heart was wounded only to those he loved and trusted most. There was a majesty in his forbearance that fairly awed me, and I often felt rebuked by it.

Such causes, together with his removal so early in life from domestic ties, and learning too soon how little true friendship the world can give, operated to make him distrust all men until they were proved true; but, once satisfied, his friendship was true and lasting as life.

His birthplace was on Lord's, now Asylum Hill, near the house of the honored, genial, and benevolent poetess of whom Connecticut is so justly proud. As a boy she loved and admired him, and amid all his wanderings by sea and by land, she seems ever to have kept him in the kindest remembrance. When pecuniary troubles had driven his father from his pleasant home on the hill, and when the inventive boy had few friends to help him in his ambitious plans, she ever spoke to him strong, encouraging words of confidence in him and his aims. This kindness to the boy was never forgotten by the man, and she had no truer friend than he. Among his papers are many notes and letters from her.

In 1848 Col. Colt returned to Hartford, his native place, and in a small three-story building on Pearl Street, all of which he then thought too much for him to occupy, he commenced the manufacture of his arms.

In 1849 he went to Europe, going as far as Constantinople, and, after his return, received from the Sultan a superb snuff-box, the lid wholly incrusted with diamonds, and the sides and bottom covered with the most exquisite enamelling. This present, so beautiful in itself, was scarcely the most appropriate one for a person who never used tobacco in any form.

Business rapidly increasing, he removed in the same year to a much larger building, built for a cotton factory, between Grove and Potter Streets.

In 1852, still increasing business prepared the way for carrying out his cherished plan of building a large Armory,—the largest private Armory in the world. He bought up a large tract of land in the South Meadows, with the design of inclosing a part of it within a dyke, for the purpose of protecting it against the spring inundation. This dyke is about one and three-quarters of a mile in length, from ten to twenty feet in height, the base about one hundred feet, narrowing to the top, where it is over forty feet wide. A portion of this lies parallel with the river. Trees were early planted on both sides the dyke, forming one of the most charming drives in the vicinity. The following verses, from the pen of Mrs. Sigourney, appeared in the *Hartford Times*, the first spring after the dyke was built:—

CONNECTICUT RIVER AND COL. COLT'S EMBANKMENT.

"Our river was swollen in its vernal pride
 By streams from the northern hills supplied,
  And snows from the mountains cold,
 And as it circled a happy shore,
 Still merrily brimming its margin o'er,
  A pleasant murmur it seemed to hold,
  And thus it talked with its eddies bold :
   'I've kept for a year
   In my quiet sphere,
 Though the last Summer's sultry sky
 Threatened to drain me bare and dry,
 Yet now, while my coffers are full and to spare,
 I'll have a good pastime and banish care.'

  'Hurrah!' quoth he,
  'For my Spring-time spree!
  Over the meadows I'll flow,
  Into the dwellings I'll go,
  Into their doors without knocking,—
 Through their best parlors I'll play ;
  The mother her babe may be rocking,
 But she'd better hasten away.
  Into ladies' chambers I'll peep—
 Into sanctums of young and old,
  I'll startle the sluggards up from their sleep
 With a dash of my waters cold.

Boats at the windows! boats at the doors!
. Toss in the children and ply the oars!
Chickens shall cackle and pigs be drowned—

ERRATA.

Page 308—read *thirty-two* feet for *twenty* feet.

To think of the mischievous pranks he'd play.

But suddenly he frowned,
  And in amazement cried,
'Who built yon giant mound,
  With its castellated pride?
Who raised its elm-tree crown?
  Ho, Wind and Rain! Ho, Storm and Tide!
Come, help me batter it down!'

So then, in a fury of rage, he drove ,
Against that dyke and its planted grove,
Night and day, with might and main,
And his allied forces, Wind and Rain,
      Day and night
      That Sebastopol height
He besieged and battled, and smote in vain.

At length, all baffled, he drew away
His marshalled hosts from the bootless fray,
Retreating sullenly and slow,
With muttered words of shame and woe:
So Xerxes turned, with fallen crest,
From conquering Greece of the dauntless breast—
For one who erst in his boyhood's hour
  Sported amid yon hillock's sheen,
Had vanquished the flood, whose beauty and power
  Were the pride of his native valleys green."

A plan so vast, so original, and imperfectly understood, encountered no little opposition from many who envied and feared him; but in spite of all obstacles interposed he pressed onward, with only slight and inevitable delays.

IIis energy, industry, and perseverance could not be suppressed, but were rather strengthened by opposition, and his works were carried on with almost miraculous rapidity. Now in Europe, now here, planning and executing almost simultaneously, he was a marvel of strength and celerity. This immense Armory was commenced early in the summer of 1854, and completed in the fall of 1855. He was in Europe in 1851, returning in February, 1852. In October of the same year he sailed for England, with men and machinery to establish his London Armory, the books of which were opened January 1st, 1853, and which was successfully continued until 1857.

During his rapid visit to England in the spring of 1854 he was examined by the Parliamentary Committee on Small Arms, on which occasion, in answer to a question of one of the members as to the auspices under which he established his manufacture of the pistol, he replied, that he worked on a patent of his own invention, and for his own profit, " and I am now proud of the results of my own exertions, and can paddle my own canoe."

In the winter of 1854 and 1855 he went to Europe, visiting Russia. He was treated with great kindness by the late Emperor Nicholas, and was received by him in his private room at the Winter Palace, St. Petersburg. He also saw the elder Grand Dukes several times. The Emperor presented him a diamond ring of great brilliancy and worth, with the crown and cipher of the Emperor in the centre on blue enamel; and he received one only a little less superb from the then Grand Duke Alexander.

Early in the year 1855, before the close of the Crimean

war, when he was about to return to America, the Russian Government being desirous of sending to this country one of her army officers for getting information to aid in the introduction of machinery into the National Armory at Toola, Col. Colt undertook the hazardous enterprise of bringing a captain of the Russian army through England to this country, as his valet. The captain, at least, did not breathe freely until, under the American flag, safe and free, he was bounding over the blue waves of the broad Atlantic.

On the 5th of June, 1856, at the Episcopal Church in Middletown, Connecticut, he was married to Elizabeth H. Jarvis, the eldest daughter of the Rev. William Jarvis, then residing there. The ceremony was performed by the venerable Bishop Brownell, who had also years before performed a like service for the parents of the bride. A large concourse of friends and distinguished strangers were gathered from many States, to offer their congratulations and good wishes on an occasion promising so much happiness. On the next day but one, he sailed with his party of four persons for Liverpool, on the steamer Baltic, Captain Comstock, landing there on the seventeenth of the same month, after a pleasant voyage. When he had spent about six months in Europe, combining business and pleasure, he was suddenly recalled to his home by the death of the secretary and treasurer of his Arms Company. Ten weeks of the summer and autumn were passed in Russia, where his friend Mr. Seymour was at that time the esteemed United States Minister at the Court of the Czar. By him, Col. Colt, and his brother-in-law, Mr. Jarvis, were made attachés of the United States Legation on the approach

of the coronation of the new Emperor, Alexander. They
were thus afforded an opportunity to be present at the
coronation scene itself, and all the balls and fêtes consequent upon the most grand and gorgeous pageant, after
presentations to their Majesties and the other members
of the Imperial family then present in St. Petersburg.
On this occasion, the great powers of England, France,
and Austria were all vying with each other, and with
Russia, in the courtliness and magnificence of the display.
Among the special ambassadors to the coronation, representing their own sovereigns, were Earl Granville of
England, Count de Morny of France, and Prince Ester-
hazy of Austria, Prince de Ligne of Belgium, &c.

After the tour of Belgium, Holland, parts of Austria,
the Tyrol, and Bavaria, a short time was given to Paris
before embarking again for America, in time to land at
New York the latter part of November.

In February, 1857, he moved into his new house, not
quite completed. It is a pleasant thing to recall now,
how that home was loved,—how each year seemed to
render it all the more dear to him, and to add to his
disinclination to go from it. A little son was born on
the twenty-fourth of February, 1857, and was named
Samuel Jarvis. For ten short months, the bright, loving
little baby made new sunshine in our happy home; but
when he had made himself so tenderly beloved, he was,
after a short but painful illness, borne so patiently,
gathered into the arms of the Good Shepherd; but our
home was very, very desolate, without the darling.

While our hearts were bleeding with the deep wound
of a first bitter bereavement, our dear friend, Mrs.
Sigourney, sent us the following lines of sympathy:—

SAMUEL JARVIS COLT.

Rest in thy bed, my darling,
  Where the bright fountain plays,
Where flowers of richest fragrance throng,
  And birds with carolled lays.

Thy little life was measured
  By moons, and not by years;
And sweetly closed before it reached
  The alphabet of tears.

Closed like a tinted sunbeam,
  That knows no shade of gloom;
Baptismal water on thy brow,
  And prayers to bless thy tomb.

Love o'er thine infant pillow
  Kept watch with steadfast mind,
And homeward took its flight with thee,
  But sorrow stayed behind.

It stayed behind, and weepeth
  Above thy beauteous clay:
Unfold thy snowy cherub's wings,
  And fan that grief away.

Or with thy harp of melody
  Enforce their glorious gain,
Who 'scape the battle here below,
  Without a wound or stain;

And in their Saviour's presence,
  Where naught can e'er annoy,
Expand in seraph power, and taste
  The plenitude of joy.

Beautiful as the most perfect sculpture, and as cold,
lay our perished lily, in the casket, as we carried him
out in the wintry blasts of December, to "sleep his
last sleep," "where the bright fountain plays," and we
returned to our childless home so desolate,—stunned by a
first great grief,—yet still striving to say from the heart,
"Thy will be done." A few weeks after the darling

40

died, Col. Colt removed his family to Washington, where
business relating to his patent-extension called him for
the winter, and where he had before taken a house, that
the child might have every home-comfort. When the
first birthday came, we were far away from the little
grave that held so many buried hopes; but our kind
friend, Mrs. Sigourney, ever mindful of those whom God
has smitten, sent us lines of Christian solace. In the
following April we returned to our dear home.

The osiers on the dyke were now well started, and
soon Col. Colt began to devise the erection of a building
for manufacturing them into useful articles, and also to
furnish employment to the families of the armorers, the
women and the children, who could with a little care
and practice soon become very expert in weaving the
pliant osiers into graceful and useful forms, and thus
earn for themselves a comfortable support. After the factory was complete, cottage after cottage rose near it, and
soon a little village, Swiss in architecture, burst into life,
adding a most picturesque feature to the quiet landscape,
and furnishing homes to those who wove the osiers into
baskets, chairs, and tables, or the various other kinds of
furniture which in time they have come to manufacture.
This was the first establishment of the kind in the
country; but now, following Col. Colt's lead in this,
as in his other and greater work, several have been
commenced in the West.

On the twenty-fourth of the succeeding November,
another little son was given,—and home grew joyous
again in the smile of the bright baby boy, named for his
two great-grandfathers, Caldwell Hart. Next came little
dark-eyed baby Lizzie, the pet of all, but especially of

her father. The first daughter—how closely she crept into that fond father's heart. She was a noble, beautiful child, with large, soft eyes, full of love and intelligence, and a queenly little form,—who would gracefully hold out her pretty dimpled hand to be kissed, as if she were indeed a little queen in the midst of her court. One of her godmothers sent her the following lines when she was just a month old:—

TO MISS ELIZABETH JARVIS COLT, JUNIOR, ON MY FIRST VISIT TO HER, MARCH 22D, 1860, SHE HAVING ATTAINED THE AGE OF ONE MONTH ON THAT DAY.

Lizzie Junior,—sweetly fair,—
Eyes of black, and auburn hair,
Tiny feet, so soft and white,
What a dainty little sprite
Art thou in thy nurse's arms,
Cheering all with infant charms.
See,—with noble, curly brow,
And his best Parisian bow,—
Wondering much to see thee here,
Collie gives thee kisses dear.
He, ere long, with brother's pride,
Shall thy timid footsteps guide
Gayly round the broad parterre,
To the green-house rich and rare,
Or through beds of fragrant pines,
Where the winter strawberry shines.

Luna once hath filled her horn,
Only once since thou wert born;
But how soon with crescent clear
Will she light the finished year,—
And how soon thy baby face
Kindle into childish grace.
So, when youth and beauty bring
Thee their brilliant offering,
Take good care, by duty led,
In thy mother's steps to tread;
And from every sorrow free
May life's changes be to thee,
Little Lizzie,—sweet and fair,
Darling of thy parents' care.

In August of that year, 1860, Col. Colt chartered a
small steamer, and, with his family and a few dear
friends, spent a most delightful week on the water,
going as far as Nantucket and Martha's Vineyard, and
touching at all points of interest along the coast. The
gentlemen found plenty of amusement in fishing. It was
a pleasant week, that none who enjoyed it will soon
forget; and when the little party separated, and each
went to his own home, in firm health and spirits, who
then would have foretold that the same little band should
never all meet again in this life!

That bright, pleasant summer, too full of happiness to
last, was followed by the saddest autumn,—for, when the
chill October winds were blowing, our little Lizzie
meekly "folded her pale hands," and closed the dear
eyes forever. Her death seemed to have struck a
death-blow at the life-springs of her father. He, who
had borne unflinchingly every ill and burden of life,
sank down before the open grave of the guileless babe,
and for weeks did not leave his room. Never will the
scene be forgotten, when his little darling, covered with
flowers, herself the purest, fairest one, was carried into
his sick-room, in the casket wherein she was laid for
burial, his agony, as he looked for the last time on the
face so lovely in death—but so still—with no smiling,
eager welcome for the father she adored. God had sorely
smitten him; yet, I often think now, He was leading him,
by a way which we knew not, to the land where those
darlings had gone. The day came all too soon, when
we must put away from our sight even the beautiful
clay of our darling; and he who had loved her so
tenderly was too ill to leave the house. From his room-

window he watched the little train of mourners, as slowly and sadly the little one was carried to sleep by her brother's side, where so often, through the sunny summer days, she had played and slept in her nurse's arms. The prayers were said,—the dust was given to dust by him who had poured the water of baptism on her fair young brow, and sealed her with the Saviour-seal. And when we turned tearfully away, the glorious October sunset was gilding all things, even those little tombs, in heavenly light, and seemed to whisper tenderly, "Their angels do always behold the face of my Father which is in Heaven."

Well I remember him on my return, sitting with the portrait of our baby before him, and convulsed with such grief as one seldom sees. The quiet peacefulness of the burial-scene had touched him deeply, and he expressed the wish that when he died he might have such a funeral,—only near and dear ones to follow him there then. Poetry wakened for him its "sadly soothing strains," but they proved more sad than soothing.

For three months after this, he was ill with that racking, excruciating disease, the gout. Yet in all this time, when the body was suffering so unspeakably, the mind and will were strong as ever. Even on his sick-bed he managed and directed the affairs of the Armory, with almost the minuteness, and all the clearness of health. What human frame could long stand such an herculean weight and strain!

When able to leave home, he sailed with his family, in February, 1861, for Cuba, hoping, in the more genial climate of the tropics, to throw off the blighting disease. This hope was partially realized, but firm health had gone forever. His little son, but little more than two years old, was a great comfort and amusement to him during his stay in Cuba.

The breaking out of the civil war, soon after his return, brought to him increased cares, and while his physical condition required entire rest, he was making arrangements for doubling the capacity of his already enormous Armory. He sorrowed deeply for the distracted state of the country, and was among the first to offer aid to the Government, when the rebels fired on the glorious old flag that had been his pride and boast, and had thrown its sheltering folds over him in so many foreign lands. He was always a democrat, and a firm supporter of the policy advocated by Douglas, who was also his warm personal friend.

A large addition to his residence was commenced. With loving care, rooms were planned for the comfort of his wife and children; but before these were finished he was called hence. His work was done, and nobly done. To human sight it seemed that there was much in the future for his guiding hand to do; but "He who doeth all things well," and who smiteth in love, judged otherwise.

His taste was refined and elegant; his judgment correct and critical. The laying out and beautifying of the grounds about his home, bringing order and harmony out of confusion and roughness, was a source of continuous delight to him. Each year he made enlargements and improvements of the area devoted to cultivation, and the outlay was repaid a hundred fold in his own enjoyment of it, and in the thought that he was making the dear home more attractive to those most beloved.

He had planted a large orchard of French dwarfed fruit-trees, and often spoke of the pleasure he anticipated in watching the children, when they could run about among them and gather the fruit with their own baby hands. There was no undertaking so vast, as to deter him from attempting; there was nothing that might give pleasure or comfort to others, too small for him to remember to perform. He never forgot a favor or a kindness, and repaid, with thoughtful appreciation of their services, those who had done their duty while in his employ. Even the animals who had served him well were cared for in their old age, and had, as he used to say of his first and favorite horse, Shamrock, "a pension for life." Poor old Shamrock outlived his master, and died of old age, in the autumn of 1863, at the age of almost forty years.

With hypocrisy he had no patience—no dealing; and his scorn of it, and his love of truth, were very prominent traits in his character. Mrs. Sigourney very justly alludes to this when she speaks

"Of the fearless truth that scorned
His frailties to disguise."

He had a quick, ready wit, which rarely failed him; and none better enjoyed or appreciated a clever joke. To my parents he was ever a kind, thoughtful, affectionate son; and to my brothers and sisters, all that a tender, true-hearted brother could be; while their grateful remembrance of his kindness, and untiring ministrations to him in his frequent attacks of illness, drew them very near to his affectionate heart.

A friend, in speaking of him after he had passed

away, was recalling the singular force of his will, his
rapid decision and prompt execution, and the magnetism
of his presence, which inspired all with confidence in his
power to lead them, said—"Had he received a military
education, he would have been one of the greatest gen-
erals the world has ever seen."

Of his faults I speak not; that must be left to other
lips—to another pen than mine; more truly than any
other he filled my ideal of a noble manhood, a princely
nature, an honest, true, warm-hearted man.   There are
spots on the sun, and if its brightness serve to make
them more visible, so do they, by contrast, but add a
more resplendent glory to its radiance.

> "No longer seek his merits to disclose,
>   Or draw his frailties from their dread abode,
> (There they alike in trembling hope repose),—
>   The bosom of his Father and his God."

On the 23d of May, 1861, Henrietta Selden, the second
daughter, was born.   Like little Lizzie, she too, at a very
early age, showed the same absorbing love for her father,
rejoicing his saddened heart with new hopes for the
future.   During the summer he had two short attacks of
gout, but recovered from both rapidly.   The last few
weeks of his life, he was much better than he had been
since the long illness of the previous winter.   Those were
very happy weeks; the children bright and well, with
almost every earthly wish gratified, how could we dream
of the fearful storm so soon to burst upon us?   At the
pleasant Christmas-time, a cold brought on what appeared
to be only a slight attack of gout, not very painful, but
confining him to his bed.   On New Year's day he was
much better, and received a call from an old friend of

his grandfather's, which he much enjoyed; he was very
cheerful, and seemed so proud to show him his boy of
three summers, and the little fairy-like baby Hetty, whose
bright sunny eyes and pretty curls won all hearts. On
the fourth of January he was sufficiently recovered to sit
up all day, and attended to a little business. Although
he was not able to walk, our hopes were very sanguine
for his speedy restoration to health. 'Twas but the lull
in the tempest, before it should return with fatal force
and desolation to lay waste the happiest home. Toward
evening I went out for a little time; on my return I
found that he had sent for a Bible, that had been given
him years since by his father, and had been reading;
and his last work on earth was to fasten into that book
a piece of poetry he had cut from a paper while at
St. Catherine's Springs the previous summer, entitled,
"What! Leave my Church of England!" It must have
touched an answering chord in his soul, as he had no
particular fondness for poetry. He had been educated in
the Congregational faith, his father having been for many
years a member of the old Centre Society; but he himself had been an attendant on the services at St. John's
(Episcopal) Church for some years.

He was looking very weary, and soon went to bed. I
sat and read to him all the evening, until quite late.
Noticing that he was much flushed, without saying any
thing, I bathed his face and brushed his beautiful curls.
Never can I forget the bright, pleased smile with which
he looked up and thanked me for "knowing just what
I wanted." Next morning he was quiet and cheerful,
and had a long conversation, in which he told me many
things that, as I now look back upon, seem to have

41

been said for my guidance in the dark hours so soon
coming, when his strong arm and kind care could aid
me no more. About one o'clock he was assisted to rise
and dress, and he enjoyed seeing the children very much
for an hour or two, when, without a warning, his mind
began to wander, even then devising a pleasant surprise
for his little son. I believe he was conscious of the
change as soon as I was, and his first thought was for
me, in the grief that he felt was coming, insisting upon
sending for our dear mother, and dictating a note for me
to write to her. As soon as possible the physician was
summoned, and he at first thought it only a temporary
thing, and administered a quieting medicine; but when,
after an hour or two, the excitement rather increased
than lessened, his friend and physician, Dr. John F.
Gray, of New York, was telegraphed to come at once.
He had the most perfect consciousness that he was losing
control over his mind, and expressed the tenderest care
and pity for me. It was a painful evening to all. Towards morning he grew quiet, and fell into a gentle
sleep, from which he awoke calm and in his right mind;
and when Dr. Gray arrived, at noon, no one could have
imagined that there had been such a cloud upon that
clear, strong mind so short a time before. Tuesday noon
the excitement came back again, and for two or three
hours he continued talking almost 'uninterruptedly and
incoherently, when gradually it all passed off and reason
resumed her sway. In this short interval of consciousness, the last, as it proved to be, in which he could
hold any conversation, the clouds were lifted, and he
looked off into the great unknown future with a calm
serenity and peace. Beautifully and touchingly he told

me that death was very near; of his trust in God's love and mercy; of his striving to do right according to his sense of right, though all things looked very differently to him now, with death so near; of his forgiveness of all who had injured him (and the God who knows how deeply he had been wronged in some shameful instances, knows, too, how much that forgiveness implied). He talked of the darling children, confiding our boy to my tenderest care and love; bidding me keep that little one, so soon to be fatherless, from all evil,—to teach him all good,—with the solemn earnestness only those can command who stand between the seen and the unseen, the living and the dead. He told me that our bright, sweet baby was going too, and did not ask me to keep her from evil, for in that solemn hour it seemed revealed to him, that the Shepherd of the lambs very soon would gather her untried soul into His own fold, to be blessed forever. Then he bade me a last farewell, calling me his faithful, loving wife, asking me to carry out all his plans so far as I might; and with his last kiss whispered that, when God willed, I should go to him beyond the grave, where partings never enter. Then reason again gave way, and, as he said himself, "it is all over now." Yet life lingered on through two more days of pain, with now and then a gleam of hope, until Friday morning, January 10th, 1862; when his great soul passed away, to the untold mysteries of the spirit-land.

On Tuesday, January 14th, at three o'clock, the funeral services took place at his late residence; hundreds gathered in the grounds and on the street, who could not find standing-room in the house. The Putnam Phalanx, of which he was a member, were present in uniform.

All the armorers, under the lead of William H. Green, one of the contractors at the Armory, and numbering fifteen hundred men, each with the badge of mourning on his arm, marched in solemn procession to the home of their late employer, to look once more upon that noble face, so full of peaceful majesty, as it lay in its last sleep.   Passing through the house, they formed in line between it and the grave, so that the body was carried to its last resting-place amid the men for whom he had done so much, and in whose well-being he had taken so deep an interest.

The Rt. Rev. Bishop Williams, and his pastor, the Rev. Dr. Washburn, conducted the services, assisted by the clergy of all the other Episcopal churches in the city.   Of those who bore him to the tomb, four were among his earliest friends; two represented the Armory, and two the Phalanx.

They were Col. Charles A. May, of New York; Judge Charles L. Woodbury, of Boston; Ex-Governor Seymour, and the Hon. H. C. Deming, of Hartford.   E. K. Root, Esq., and Mr. H. Lord, Mr. J. H. Ashmead, and Mr. E. D. Tiffany, also of Hartford.   And there we left him, near the little ones so dearly beloved.

Hartford knew on that day that she had lost one of her noblest sons.   The flags were all displayed at half-mast on the day he died.   In many a home his prosperity had aided there were mourning hearts; and all felt then, that he who that day had fallen was among those whose place could never be filled.

During his last illness he kindly remembered those who were faithful in the performance of their duties, and by my hand prepared New Year's gifts for several in his employ. How little I then thought that it was the farewell,—the gift to tell them, when he should be seen no more, how he had trusted and thanked them.

On the day succeeding the funeral, the following note and stanzas were sent me by our valued friend, Mrs. Sigourney, who mourned for him as a son:—

HARTFORD, *January* 15, 1862.

MY DEAR MRS. COLT:—My heart is so much with you this morning, that I must indulge in a few words of sympathy. Heavenly consolations, I have no doubt, sustained you yesterday, through that trying and solemn scene. I never saw so imposing a funeral, where there was such deep sorrow on the face of every attendant. That of a prince might have had more pageantry, but here the tears of the sons of toil attested the heart's grief for the loss of their friend and benefactor, which was touching to every beholder. I hope you are comfortably well to-day, and the dear little ones, and that Collie has entirely recovered from the indisposition he had last week. I am rejoiced that you are surrounded by so many kind friends to comfort you in the loss of one of the most affectionate hearts ever vouchsafed to man, and whose love for you was so confiding and perfect. My love to them all; and will you say to Mrs. R., I wish she might be able to call before she leaves, as we both made ineffectual attempts to see each other during her former visit to you.

Will you, my loved friend, accept the poetical tribute I inclose, as a mark of gratitude to his memory, and of sympathy in your grief, from yours,

With true affection,

LYDIA HUNTLEY SIGOURNEY.

ON THE DEATH OF SAMUEL COLT.

And hath he gone, whom late we saw
  In manly vigor bold?
That stately form and noble face
  Shall we no more behold?
Not now of the renown we speak
  That gathers round his name,
For other climes beside our own
  Bear witness to his fame;

Nor of the high inventive power
  That stretched from zone to zone,
And 'neath the pathless ocean wrought,
  For these to all are known;
Nor of the love his liberal soul
  His native city bore,
For she hath monuments of this
  Till memory is no more;

Nor of the self-reliant force
  By which his way he told,
Nor of the Midas-touch that turned
  All enterprise to gold—
And made the indignant River yield
  Unto the osiered plain—
For these would ask a wider range
  Than waits the lyric strain;—

But choose those unobtrusive traits
  That dawned with influence mild,
When in his noble mother's arms
  We saw the noble child,
And noted, mid the changeful scenes
  Of boyhood's sports or strife,
That quiet, firm, and ruling mind
  Which marked advancing life.

So, onward as he held his course
  Through hardship to renown,
He kept fresh sympathy for those
  Who cope with fortune's frown—
The kind regard for honest toil,
  The joy to see it rise,
The fearless truth that never sought
  His frailties to disguise—

The lofty mind that all alone
　Gigantic plans sustained,
Yet turned unboastfully away
　From fame and honors gained;
The tender love for her who blest
　His home with angel-care,
And for the infant buds that rose
　In opening beauty fair.

Deep in the heart whence flows this lay
　Is many a grateful trace
Of friendship's warm and earnest deed
　Which naught can e'er replace;
For in the glory of his prime
　The pulse forsakes his breast,
And by his buried little ones
　He lays him down to rest.

And thousands stand with drooping brow
　Beside his open grave,
To whose industrious, faithful hands
　The daily bread he gave—
The daily bread that wife and babe
　Or aged parent cheered,
Beneath the pleasant cottage roofs
　Which he for them had reared.

There's mourning in the princely halls
　So late with gladness gay—
A tear within the heart of love
　That will not dry away—
A sense of loss on all around,
　A sigh of grief and pain:
"The like of him we lose to-day,
　We may not see again."

Scarcely was he hidden from our sight when the babe of my bosom began to droop like a smitten flower,—while she daily talked to, and caressed with tender, touching devotion, the picture of the loving father she so much missed. Two days she suffered, and then, with outstretched hands, asked mutely to be taken once more to my arms, and nestled so lovingly to my almost break-ing heart for the last time, to die there. Ten days had taken from me husband and babe; and two weeks from the day he died, a little mourning band followed the blighted flower to the father's tomb. The stone had been rolled away, and the little one slept with her father; the oak and the tiny flower rest in one grave, till the angel shall sound the last trump, and the dead shall be raised incorruptible and full of glory.

LINES ON THE DEATH OF HENRIETTA SELDEN COLT.

" A tomb for thee, my babe!—
Dove of my bosom, can it be?—
But yesterday, in all thy charms,
Laughing and leaping in my arms—
A tomb and shroud for thee!

" A couch for thee, mine own,
Beneath the frost and snow!
So fondly cradled, soft and warm,
And sheltered from each breath of storm—
A wintry couch for thee!

" Thy noble father's there—
But the last week he died!
He would have stretched his guarding arm
To shelter thee from every harm—
Nestle thee to his side.

" Thy brother slumbereth there,
Our first-born joy was he—
Thy little sister, sweetly fair,
Most like a blessed bird of air,
A goodly company.

" Only one left with me—
One here and three above.
Be not afraid, my precious child,
The Shepherd of the Lambs is mild—
Sleep in His love.

" Thou never saw'st our Spring
Unfold the blossoms gay;
But thou shalt see perennial bowers,
Inwreathed with bright and glorious flowers,
That cannot fade away.

" And thou shalt join the song
   That happy cherubs pour
In their adoring harmonies;
I'll hear ye, darlings, when I rise
   To that celestial shore.

" Yes!—there's a Saviour dear—
   Keep down, O tears that swell!
A righteous God who reigns above,
Whose darkest ways are truth and love—
   He doeth all things well. "

Early and very swiftly the evening shadows closed in upon the noontide brightness of the happiest heart-life, quenching, in darkness and tempest, the brightest hopes, the fairest realities. What wonder if sometimes the tortured heart forgets that the Father's hand chasteneth those whom He loves, or that Faith's snowy wings sometimes droop wearily above the graves of the brave, the true, and the beautiful. But, though he has gone from us, he yet lives,—lives in his great work, the monument which his own faith and energy built; in many a home that he has made happy and comfortable; in the true hearts that mourn so sadly for the breaking of "the strong staff and the beautiful rod,"—lives even in the tiniest flower his taste has planted for those that should come after. And when the sick and the dying are cheered and refreshed by the fruits of the trees his hand and thoughtful kindness have planted, in all these, he, "being dead, yet speaketh," in trumpet-tones, of faith, and hope, and triumph. God grant, in His mercy, that the broken ties may be again united, where there is neither temptation nor parting, where tears shall be wiped from all eyes, and the weary shall find eternal rest.

42

"If I forget *thee* for a while.
Then, like some mournful strain.
   Thine image seems to chide my smile,
And o'er me comes again,—
   O'er each still hour it comes from far,
With thoughts of by-gone years,
   Reflected like a heavenly star,
In the deep fount of tears.
   That fount of tears it hidden lies,
Within my Saviour's breast,
   And I will leave thee in the skies,
And that deep fount to rest.
   O Thou who knowest our secret frame,
And every inmost grief,
   In Thee I leave that long-loved name,
And find in Thee relief."

ARMSMEAR, *Nov.* 10, 1864.

My Dear Mr. Barnard:—You ask if you may print
the notes I gathered into a connected narrative, embracing the chief events in my husband's life, for your assistance in the Memorial you were intending to prepare?
It was written, as you know, for your eye and aid alone.
My aim was, to show some of those phases of his mind
and heart of which the world knows little or nothing.
His unflinching courage and resolution, his perseverance,
his belief in himself, and in his own power to conquer
obstacles that would have deterred and conquered a spirit
less strong and firm, were seen and felt by all. But few,
comparatively, knew him as he was at home, where the
active, untiring man of business was laid all aside, and
he became only the kind, thoughtful friend, the wise
adviser, the loving, devoted father and husband; or that,
with his gigantic strength of will, was joined a heart
most kind, gentle, and affectionate to those he loved and
trusted; or how, with his impulsive, passionate tempera-
ment, he yet bore wrongs that would have made the
meekest unforgiving, without one word of unkindness or
reproach.

If you think that the simple words prompted by the
affectionate admiration of a bereaved heart, can tell the
story better than " the pen of the ready writer," you are
at liberty to use them as you please.

Since those pages were written, many a change has
come over the scene and the actors therein alluded
to. The grand old Armory, destroyed by (as I believe)
incendiary fire, wiped out in little more than one short

hour the gathered excellence of many years. Yet that event brought out in strong colors the noble devotion of the armorers. Those men, some of whom had been with it from its rise, wept, as for a fallen friend, over its ashes. In alluding to the fire, please do not omit a handsome tribute to the brave, true men, who risked their lives to save the New Armory. They conquered when the steamers failed. God bless them.

Mr. Root, who was a valued and esteemed friend, and, after my husband's death, the President of the Arms Company, has also laid down his work and followed to the land where the weary rest. And you will, I know, deem it only a fitting tribute to a worthy life, to devote a page or two to his memory, in connection with the business relations between my husband and himself for many years, and the warm personal interest of Mr. Root in all relating to the Armory.

Col. Charles A. May, another of his dear and life-long friends, the brave soldier, who on the bloody fields of Mexico gained deathless laurels, the genial companion, the high-toned man, he too fell asleep on the 24th of December, 1864.

Later still, on the 10th of June, 1865, Mrs. Lydia Huntley Sigourney, an endeared friend, the sweet poetess, whose gentle strains have soothed the heart of many a mourner, the truest of friends, the humblest, most consistent of Christians, passed away to the reward of the pure in heart.

Nor can I fail to link in this little circlet of tried friends the name of the dearly loved brother, John S. Jarvis, so recently and suddenly cut down in the midst of his bright youth. From boyhood he was a great favorite of my kind husband's, his mechanical taste and skill, as well as the noble, generous qualities of his heart, drawing them very near together. His hand has aided in preparing some of the illustrations of the memorial volume we have thought of so long, and which he so ardently desired to see completed. Now it can never meet his eye, but God grant that he beholds fairer scenes than earthly artist paints, and that they who so tenderly loved each other here, are met in the light of that morning star that never fades or wanes :—

> "There shall we meet, parent and child, and dearer
> That earthly love, which makes half heaven of home;
> There shall we find our treasures all awaiting,
> Where change and death and parting never come;
> There, brother, may we meet and rest,
> Amid the holy and the blest."

I have written you a long note in reply to your question, but it seemed impossible to express all I desired to in few words. But I must beg you to accept my warm thanks for the interest you have taken in this Memorial of my departed husband, and the assurance I feel that but for you it would not have been so pleasingly carried out.

My kindest wishes will follow you to your new home, that your efforts there may be rewarded and blessed a hundred fold.

And believe me to be, with high esteem, your very sincere friend,

ELIZABETH H. COLT.

ARMSMEAR, *Sept.* 8, 1866.

CORRESPONDENCE.

CORRESPONDENCE.

WATERVLIET, N. Y., *May* 31, 1837.

DEAR SIR:—Immediately after the adjournment of our experimental board, my public duty will take me to the far West. Having none other than holster pistols, and wishing to have a convenient and ready weapon about my person, I have to request you to furnish me one, or a pair, if not too expensive, of your pistols, to be subjected to your own proof, &c. Please write me at Watervliet, if you can oblige me, when, and *at what price.* I shall probably want them by the Fourth of July.

Yours respectfully,

W. J. WORTH.

WASHINGTON CITY, *July* 30, 1840.

SIR:—I have the honor to address you on the subject of arms for my company in the Florida war, and permit me to plead, in extenuation of my presumption, an ardent desire to contribute my mite toward achieving a consummation of the efforts of Government in that particular.

43

The company of which I am captain is Company A of the Seventh Regiment of Infantry, commonly called grenadier, or the right flank company of the battalion. As this company is ordinarily the leading one, and is apt to be first in action, and perhaps have to sustain the whole fire of the enemy, while the other companies are falling into line,—upon whose efficiency of fire, whether with civilized or savage foe, much must depend,—I have the honor to request, if in any way it can be permitted, that Colt's repeating fire-arms be furnished the men, this being a weapon with whose merits I have been thoroughly made acquainted, after the use of one myself for more than a year, and with which no arms I have seen (the musket not excepted) bear a comparison.

Long since, when at Fort King, East Florida, I applied at the ordnance dépôt at Garey's Ferry, Black Creek, for these arms for my company, and was informed that there were none on hand ; for, having satisfied myself that the simplicity of construction, which is but a little more than a system of levers, surpassed that of the ordinary gun-lock, and could be easier repaired, while its efficiency as a small weapon was unequalled by any in service, I was anxious for its use in my company, and considered thereby that my command would be equivalent to a much superior force.

It is true, we have needed men, perhaps, more than any thing else, during the past year, to fill up the ranks of our companies ; but, were the soldiers supplied with this arm, it would contribute more even than numbers to a successful issue in our combats. Indians in superior force will readily attack our smaller parties, but are wary in coming in contact with equal numbers ; and such is the nature of the country, that to hide when they see a larger body of soldiers is an easy matter: witness the great number of cabins burned and fields destroyed, and so few of the enemy taken or killed.

This weapon—eight times as efficient in the fire as the musket—would inspire confidence in our own ranks, however great the disparity otherwise, when we meet in action with the foe, and the after-trial. I condemn unequivocally Hall's carbine, and prefer the old arms to every other patent, except this of Colt's, that I have seen. I must recommend this strongly, as that to which in the hour of danger I would be most willing to trust my reputation and life, and as being peculiarly desirable for my company, on account of its relative position in the column and line of the battalion.

In conclusion, permit me to say that a bayonet would be an advantageous appendage to this arm; and the lever rammer, which can be easily prefixed, would add to its efficiency.

Again apologizing for this intrusion upon your time, I have the honor to be, very respectfully,

Your most obedient servant,

G. T. RAINS,

Captain Seventh Infantry, commanding Company A.

P. S.—Should the honorable Secretary of War consider the above as premature, will he please to bear this application in mind should the weapon be adopted into service at any future period?

Respectfully,

G. T. RAINS,

Captain.

WAR DEPARTMENT, *August* 20, 1840.

SIR:—I received in due time your letter of the 30th ultimo, recommending that Colt's repeating arms be furnished to your company, and would have made an earlier acknowledgment of its receipt, but for a great press of business. I now have to inform you, that it is not deemed expedient to comply with your wishes at present, and that your desire to have the application borne in mind, in the event of the weapon being adopted into the service at any future period, is acceded to.

Very respectfully, your most obedient servant,

J. R. POINSETT.

HEAD-QUARTERS ARMY OF OCCUPATION,
CAMP NEAR MONTEREY, *August* 17, 1847.

SIR:—Your letter of June 7th, and the accompanying box, containing a pair of your new-modelled repeating pistols, have duly reached me. I have been much pleased with an examination which I have made of the latter, and feel satisfied that, under all circumstances, they may be safely relied upon. Be pleased to accept my thanks for this valuable present, and my best wishes for your success.

I am, Sir, very respectfully,

Your obedient servant,

Z. TAYLOR,

Major-General United States Army.

COLONEL HARNEY TO SAMUEL COLT, ESQ.

WASHINGTON, *January* 14, 1848.

DEAR SIR:—In answer to your inquiries as to my present opinion of your patent revolving fire-arms, I am free to say, that, after an experience of more than ten years in their use, commencing with the Florida war, and continuing through our contest with Mexico up to this time, no arm, in my judgment, ever yet constructed can equal them. They are, with fair usage, as little liable to get out of order as any other, and they are at least three times as effective. I consider them invaluable in the present state of affairs in Mexico, where our cavalry are always compelled to face an overwhelming force of the enemy; and they will be equally valuable on our frontiers: when it is known that we possess such arms, they will be less apt to commence hostilities with us. I also consider them, for these services, much *the most economical arms* that can be used. It is my intention to apply to the honorable Secretary of War, to have all the regular cavalry armed with your improved holster pistol; for I am confident that, if they were armed in this way, incalculable advantages would be gained.

In haste, your obedient servant,

W. S. HARNEY,
Colonel Second United States Dragoons.

MAJOR G. T. HOWARD AND CAPTAIN SUTTON TO COL. COLT.

WASHINGTON, D. C., *February* 26, 1850.

SIR:—Being requested to give our opinions relative to the merits of your justly celebrated repeating pistols, we do so with much pleasure, as we know their efficiency from long experience, and deem them to be the greatest improvement in small arms of the age.

We have been familiar with the use of your effective invention on the frontiers of Texas since 1839, and we do unhesitatingly affirm that they give the combatant greater confidence and spirit of defiance in those hand-to-hand struggles with the prairie Indians than any other arm now in use. Those prairie tribes ride with boldness and wonderful skill, and are, perhaps, unsurpassed as irregular cavalry. They are so dexterous in the use of the bow, that a single Indian, at full speed, is capable of keeping an arrow constantly in the air between himself and the enemy; therefore, to encounter such an expert antagonist with any certainty of doing good execution, requires an impetuous charge, skilful horsemanship, and a rapid discharge of shots, such as can only be delivered with your six-shooters.

They are the only weapon which has enabled the experienced frontiersman to defeat the mounted Indian, in his own peculiar mode of warfare. In those encounters which, though soon over, require a steady nerve, the greatest possible precision and celerity of movement, there is no time to reload fire-arms, even were it possible to do so, and manage your horse, in the midst of a quick and wily enemy, ever on the watch, and ready to lance the first man who may lose the least control of his animal.

In this description of service, the revolver asserts its great and unquestionable superiority over all other weapons. We state, and with entire assurance of the fact, that your six-shooter is the arm which has rendered the name of Texas Ranger a check and terror to the hostile bands of our frontier Indians. With the citizens of Texas they are considered unparalleled in the force and accuracy of their shooting, and are esteemed an invaluable weapon in offensive operations against those marauding tribes which depredate upon our frontier settlements.

As for the difficulty of keeping those pistols in order, which is sometimes raised as an objection, we have never discovered any. Those instruments can only become unfit for service in the hands of men who never possessed the proper pride of a soldier, or where discipline must be most egregiously neglected.

With great respect,

Your obedient servants,

G. T. HOWARD,

Late Major Texas Regiment.

I. S. SUTTON,

Late Captain Texas Rangers.

LORD JOHN RUSSELL TO COL. COLT.

DOWNING STREET, *December* 23, 1851.

SIR :—I am much obliged to you for the offer of a specimen of your revolving fire-arms. I shall be glad to accept one, as a curiosity, and a proof of your ingenuity.

I remain your obedient servant,

RUSSELL.

LORD PALMERSTON TO COL. COLT.

BROADLANDS, *January* 6, 1852.

SIR :—I have many apologies to make to you for having so long delayed, amidst a heavy pressure of business, to thank you, as I now beg to do, for the beautiful revolver which you have been so good as to send me. It is indeed an admirable specimen of ingenuity of invention, and of perfection of finish. I shall preserve it not only as a valuable memorial of that great exhibition which brought the nations of the world into such friendly contact, but as an evidence of that kindly feeling on the part of our brethren in the United States, to the maintenance and increase of which I attach such great importance; and I trust that for many a long year to come the people of our two countries may know nothing of each other's weapons except through such acts of courtesy as that for which I am now making my acknowledgments.

I hope on my return to town to have the pleasure of making your acquaintance, and of repeating my thanks to you in person.

I have the honor to be, Sir, yours faithfully,

PALMERSTON.

COL. COLT TO RT. REV. T. C. BROWNELL.

HARTFORD, *April* 1, 1852.

REV. SIR:—Learning from your son this morning that there had been a recent attempt to despoil your household of valuable jewels, I take the liberty of sending you a copy of my latest work on "Moral Reform," trusting that, in the event of further depredations being attempted, the perpetrators may experience a feeling effect of the great moral, influence of my work.

Although yours is a profession not generally using weapons of this description, yet (as a carnal sword once cut off a carnal ear) should occasion require its use, I trust this weapon may in your hands be instrumental in protecting your life from harm and your property from spoliation. Such at least is the sincere wish of,

Reverend Sir, faithfully,

Your obedient servant,

S. COLT.

RT. REV. T. C. BROWNELL TO COL. COLT.

HARTFORD, *April* 2, 1852.

MY DEAR COLONEL:—If it be true that "a preparation for War is the best preservative of Peace," your ingenious work on "Moral Reform" will be productive of great benefit to the world. Please accept my most cordial thanks for the beautiful copy of the work presented to me last evening. You will confer an additional obligation by calling on us occasionally, to give some needful instructions in regard to the application of its principles.

Very truly, your obedient servant,

T. C. BROWNELL.

THOMAS ADDIS EMMET TO COL. COLT.

NEW YORK, *May* 28, 1852.

MY DEAR SIR:—The pistol which you forwarded me by express came safely to hand yesterday.

I thank you for your kindness in remembering me, and I accept it in the same spirit with which it has been sent, as an offering to "auld lang syne," and as such it shall never be parted with.

It did not, however, require such a splendid specimen of the perfection of your revolver to remind me frequently of you. I confess that a more sordid feeling, to

44

wit, the conviction of how much I and those associated with us in the Patent Arms Company *have lost in a pecuniary point of view*, often *revolves* in my mind. I was always a firm believer in the utility and ultimate success of your arms, and used all my endeavors to adjust the unfortunate disagreements which severed us asunder, but it was of no avail; and while I regret *my own loss*, I most sincerely wish that you may reap the full and rich reward which so valuable an invention deserves, and live long to enjoy it.

With great respect, I remain yours, &c.,

THOMAS ADDIS EMMET.

MRS. SIGOURNEY TO COL. COLT.

HARTFORD, *September* 29, 1852.

MY DEAR SIR:—Many thanks for your kind note, for the splendid flowers, the delightful drive with the winged horses, and other countless favors. I regret not seeing you again, and that you could not have accepted my poor hospitalities for an evening, which I would have done all in my power to render agreeable to yourself and your friends, but know it is impossible. I send a letter and small parcel for Mrs. S. C. Hall, which I hope may not give you trouble. I also ask your acceptance of a book, hoping through its aid to adhere to your remembrance, and thinking you might like to cast your eye over it on the voyage, as it sketches some scenery which we have mutually visited.

But that panorama on the brow of Rocky Hill will be vivid and beautiful in my memory until we meet again. I shall trust to have a few lines from you when you are at leisure and settled in the good old mother-land.

It is wisest that parting words should be few; so I will only say that I wish you health and success in your patriotic enterprise, and good treatment from old Neptune, and a safe return, when, if I am still in life, you will receive a true welcome,

From your grateful friend,

L. H. SIGOURNEY.

COL. COLT TO KOSSUTH.

LONDON, *March* 20, 1853.

DEAR SIR:—Permit me to present you with a box containing specimens of my Patent Repeating Fire-arms, in token of my high regard and esteem.

The arms were made for you at my manufactory in Hartford, Connecticut, when you were travelling in America, and would have been presented to you by a committee of my workmen, had it been our good fortune to have seen you at Hartford, in your hasty tour through New England.

I trust, however, that even at this late day they will not prove unacceptable, and will be received with the same feeling of good-will with which they are sent. Permit me also to say, that while it would have been the greatest satisfaction to me to have shown you my manufactory of fire-arms in Hartford, it will not be less so to have you and any of your friends examine my smaller establishment, just started, at Thames Bank, Pimlico, London, at any time it is your pleasure to fix for that purpose.

Believe me, in the mean time, very respectfully,

Your Excellency's most obedient servant,

SAM. COLT.

LONDON, *March* 27, 1853.

DEAR SIR:—You have very agreeably surprised me by
your splendid pair of revolvers, which I accept with
particular gratification, and beg leave to return you my
sincere thanks for your valuable present, as well as for
the accompanying obliging lines. I was very glad to
learn that the practical superiority of your invention has
met already such well-deserved appreciation with the
Government of the United States. This is equally creditable to the merits of your genius, and to the progressive
spirit which seems to be a peculiar privilege, and a
glorious one indeed, of democratic institutions.

To me your magnificent present, dear Colonel, has a
triple value: that of your valuable sympathy, that of
artistic-mechanical merit, and that of practical use; since,
without indulging in any sanguine fascinations, I dare
hope soon to have occasion to use it, and use it I shall
with the conscientious conviction that your genius never
could have aided a better cause. From what I know of
military art, I confidently believe that the improvements
in fire-arms are likely soon to change essentially the
adopted tactics; and as the anticipation of my future
duties cannot but increase my interest in matters to which
you have so successfully devoted your genius, I am
thankful for your inviting me to visit your London
establishment, and will avail myself of it at the earliest
opportunity. Accept once more, dear Colonel, my sincere
thanks, and believe me to be, with distinguished regards
and particular consideration,

Your obedient servant,   L. KOSSUTH.

HARTFORD, *May* 12, 1853.

I trust your health continues good. We hear often of you through prints, both English and American, and in such high terms as are cheering to the heart of your early friend, who distinctly remembers your boyhood, and the glee with which you used to lead the Hill sports on Saturday afternoon. Did you get my poor "Faded Hope," and find a moment to glance over it? When shall we see you again? But it is time to close this gossiping letter. Desiring kind messages to any in the good mother-land who may chance to remember me, I remain,

Very truly and gratefully, your friend,

L. H. SIGOURNEY.

LONDON, *May* 27, 1853.

MY DEAR MRS. SIGOURNEY:—I have again to-day received a very kind letter from you, and must confess it has made me feel very culpable as well as homesick. Do not judge from my neglect to answer your former kind letter that I am entirely callous to those finer feelings of the heart which so abound in yours. You have yourself made for me the only excuse I can offer for my apparent want of feeling and neglect of my home friends. Constant application to business has almost entirely shut out private correspondence. To do my business thoroughly is, I must confess, a passion with me. Yet I have constantly hoped, and still do so, that the time is near at hand when I may so far be relieved from it, as to be able to comply with the requirements of the courtesies and refinements of social life. Pardon my long silence and the brevity of this. I am compelled to write so hastily, that this can only show you that I am not entirely unmindful of the early friend of my beloved mother, and the affectionate adviser of my dear sisters. I have seen your friends, Mr. and Mrs. Hall, but once since my arrival in London, and then spent a day and night at their charming residence in the country very happily. Since then I have been too much absorbed in business to repeat the visit. I have now before me a note begging me to call on them before I return to America, which I most certainly shall do, if it is possible, and shall tender my services for any commissions they may have for you or any other friends in America.

I shall start for Dublin this evening, and must hastily close for the post; but I promise you an early visit on my return to Hartford, next month, when I hope to find you happy and in your usual good health.

Meantime, I remain, as ever, truly yours,

SAM. COLT.

MRS. SIGOURNEY TO COL. COLT.

HARTFORD, *May* 14, 1855.

MY DEAR FRIEND:—As I take pleasure in occasionally driving with strangers who call on me about your magnificent embankment, I think I will sometimes avail myself of your proffered kindness, and be indebted for the use of some of your twelve horses during my absence.

I feel as if I had never sufficiently thanked you for your many acts of attention and disinterested goodness, among which I number your kind parting call of the last evening. Indeed, you have never known how much good they have done me, or what strength I derived in past times, by conversing with and consulting with you. Therefore, I would now fully thank you for all. It is not the majesty of intellect which overcomes obstacles, which I chiefly admire in you, but the heart. Greatness forces men to honor it, but those who have been led to know the worth of the better feelings of our nature, always discover them where they exist; and this appreciation is better than the blast of fame. I have long discovered them in you, and been the object of them,—for you have extended to me a noble regard, in return for the affection cherished for your departed sisters when under my care, and the friendship felt in past years for your parents. And now I find myself relying in my loneliness much upon you. So I pray you, as I said last evening, with a kind of maternal frankness, which I hope you will forgive, to run no unnecessary risks,—and in your periods of recreation to do nothing that can injure your health. Keep yourself well in body and mind, for *my sake, will you?*

I hope you will be yet a model of happiness in domestic life, as you have already been of mental greatness. What I say comes from the heart, so I trust it will not offend you. The Lord bless you, and bring you back in safety, to

Your grateful and attached friend,

L. H. SIGOURNEY.

C. G., *December* 21, 1853.

MY DEAR SIR:—I am much obliged to you for your tickets of admission to your interesting manufactory, and I shall not fail to avail myself of them on my return to London after Christmas.

My dear Sir, yours faithfully,

PALMERSTON.

ORDNANCE OFFICE, *February* 17, 1854.

MY DEAR COLONEL:—His Royal Highness Prince Albert proposes to visit your manufactory this day at three o'clock.

It occurs to me that you will like to be informed of His Royal Highness' intention.

Yours, faithfully,

RAGLAN.

BEFORE SEBASTOPOL, *August* 21, 1855.

DEAR SIR:—Since I wrote you by last mail I have received the inclosed from Gen. Pelissier, which I have much pleasure in forwarding to you.

I am, yours truly,

NEWCASTLE.

GEN. PELISSIER TO COL. COLT.

GENERAL QUARTERS, *August* 20, 1855.

COLONEL:—His Grace the Duke of Newcastle has had
the goodness to send me the two revolving pistols which
you had the kindness to offer me.

I accept your present, and thank you for having sent
these beautiful arms, to which you have given so ingenious a fabrication and a remarkable perfection.

They have been exceedingly admired by those of our
officers who use fire-arms (*or are interested in fire-arms*).

I am much touched at this cordial politeness, and pray
you to believe in my distinguished consideration.

PELISSIER,
The General-in-Chief.

[Translation.]   *By A. Deussin.*

SECRETARY OF CONNECTICUT HISTORICAL SOCIETY TO COL. COLT.

HARTFORD, *December* 19, 1855.

At a meeting of the Connecticut Historical Society,—

Resolved, That the thanks of this Society be, and
hereby are, presented to Col. Samuel Colt, for his present
of one of his own highly beautiful Repeating Pistols,
with all its appropriate accompaniments. As a specimen
of the inventive genius of a son of Connecticut, uniting
as it does superior grace of construction with unrivalled
effectiveness as a weapon of war, it is highly prized, and
shall be carefully preserved within the custody to which
he has consigned it.

The Society cordially unites with Col. Colt in the

45

hope that remarkable specimens of "industrial art" may be hereafter added to its collections, and may subserve, though they be "*warlike*" in their design, the "*peaceful*" purposes of its own, and of the organization of all civilized society.

Resolved, That the Secretary of the Society transmit a copy of this resolution to Col. Colt.

Attest,   CHARLES HOSMER,
Recording Secretary.

(A true copy.)

COMMODORE PERRY TO COL. COLT.

NEW YORK, April 25, 1857.

MY DEAR COLONEL:—Your favor, addressed to me at Washington, was forwarded to me at this place, and I thank you for the information it contained.

By the last mail from India I received a friendly communication from the Second King of Siam, in which he thus speaks of your revolver:

"The beautiful pistol which you had the honor of forwarding to me has afforded me much satisfaction and amusement. I was so much pleased with it, that I have taken special pains to secure as many varieties of the revolver as I could obtain, but none have given me the satisfaction of the pistol you have sent me."

The above, written in English by the king himself, is only a portion of his letter, which I will show you when I see you again.

The name of the king is S. Phra. Piu. Klau. chau. qu. hua. He is well educated and a great lover of the arts. I shall publish his entire letter in the appendix to my second volume.

I shall send him by the frigate Mississippi, who is to be commanded by a friend of mine, a copy of my work, and if you desire to present him a pair of your miniature pistols, in a pretty case, I will forward the gift with your letter. Those packages can be sent to Siam through a friend at Singapore.

The emperor and several of the princes of Japan, the governor of Shanghai, the king of Lew Chew, and the king of Siam have received your invention through me.

But in regard to the fruit of your last *and most valuable invention.* I hope the young Colonel is well, and though he has probably not yet learnt to read, I send him a book to be in readiness when he does arrive at that accomplishment.

I hope you have already received the Japanese swords, lances, &c., from the Patent Office. They certainly do not belong to the Government, and if you have not yet got them in your possession, Mr. Toucey surely will not fail to have them sent to you.

Have you yet commenced your museum of arms?

Most truly yours,

M. C. PERRY.

THE SECOND KING OF SIAM TO COL. COLT.

BANGKOK, SIAM, *May* 27, 1858.

DEAR SIR:—I know not how to express my thanks for your flattering letter and presents. I have always entertained a high respect for your creative talent, so pre-eminently displayed in that formidable weapon, your revolver. The entire series of your arms have strengthened my high

regard for your unparalleled inventions. Please accept a
few specimens of Siamese manufacture, which it affords
me much pleasure to forward as a tribute to your genius
and worth. I would only be too happy were it in my
power to augment your already well-earned and well-
deserved world-wide reputation.

Allow me to suggest, that it is a peculiarity of our
kingdom to have a first and a second king, both of whom
are entitled to the address, "Majesty." I make the
suggestion simply in the hope of preventing misconcep-
tions abroad.

Yours affectionately,

S. P. PAWARENDRRAMESR,

Second King of Siam, &c., &c., &c.

THE MAJOR KING OF SIAM TO COL. COLT.

BANGKOK, SIAM, *January* 24, 1859.

SIR:—I have great pleasure in the receipt of your kind
present of powerful and beautiful fire-arms, being our
revolving rifle and four pistols of various size, kept in
good order in the box covered with crystal, through the
introduction and care of Commodore Perry, who is now,
I have heard, dead.

Whereas Reverend Mr. S. Matlen, the American offi-
ciating consul, will return to his home, I have liberty to
intrust to his care a golden snuff-box, a silver water-pot,
a silver plate with stand, and silver cigar-case, richly gilt,
for your acceptance. I hope they will be the token of
my remembrance; and I beg to express my very heartful

thanks for your kindness toward me, and beg to present to you one of my *cards,* out of respect to you.

I have the honor to be,

Your good friend,

SOMDET PHRA

PARAMENDR

MAHA MONG KUT,

King since 2812 days ago.

P. S. I will write you again, on another opportunity, regarding your manufactures.

THE MINISTER OF THE CZAR'S HOUSEHOLD TO COL. COLT.

ST. PETERSBURG, *March* 4, 1859.

SIR:—His Majesty, the Emperor of all the Russias, my august sovereign, is pleased to present to you a snuff-box, ornamented with his cipher and enriched with diamonds, in return for the different arms of which you made homage to his Majesty, as well as to their Imperial Highnesses, the Grand Dukes of Russia.

Will you have the kindness to acknowledge the reception of the Imperial casket mentioned, which I hasten to transmit to you herewith.

Receive, Sir, the expression of my civilities.

COUNT M. DE ADLERBERG,

The Minister of the Household of the Emperor.

LIEUT. HANS BUSK TO COL. COLT.

LONDON, *April* 28, 1859.

MY DEAR COLONEL:—Allow me to acknowledge, as I do with the greatest pleasure and sincerest thanks, the receipt of your very kind and handsome present, which

Mr. Dennet handed to me yesterday. Believe me, I shall
ever prize it as a memento of the valued friendship of
the donor, and of the many pleasant hours spent in his
society, no less than for the talent and ingenuity displayed in perfecting by far the most complete specimen
of a soldier's firelock that has yet been produced. I
have the credit of knowing something about arms of this
description; and, having tried every variety of American,
as well as every European rifle, that has been produced
of late years,—and, in so doing, having fired more than
sixty-eight thousand rounds from my own shoulder,—my
opinion on such matters is, perhaps, worth more than
the mere empty praise of a "green hand." I have this
morning been trying your rifle, and inclose a diagram,
showing the result of forty-eight shots (twenty-four from
the shoulder and twenty-four from a rest), at four hundred
yards. The practice shows that, with a tolerably steady
hand and arm, your rifle does all that can be required
of such an arm. Let any one who wants to know what
a "Colt" can do, take my word, that, for efficiency and
strength of shooting, NOTHING CAN BEAT IT.

By the news which goes out by this mail, it would
seem that Europe is about entering upon a long and san-
guinary war. Had we but forty thousand stout volunteers
armed with your rifle, we might set France and Russia
at defiance; but, as it is, with the sad lot of incapables
at the heads of departments, who bungle matters usually
so miserably, we shall, I dare say, make a mess of it,—at
first at least,—as we did in the Crimea. Most likely, before many months are over, we shall have more than one
powerful enemy eager for the spoliation of this little isle;
but, thank God! we have some sound hearts, and strong

arms, and provident men, who place their trust in their
"Colt," as I do implicitly—and we keep our powder
dry!

May we hope to see you over again amongst us this
year? Come and see *what is left of us* after the French
have done their worst.

Pray make my very best remembrances acceptable to
Mrs. Colt, and believe me, as ever, my dear Colonel,

Your most sincere friend and well-wisher,

HANS BUSK,
First Lieutenant V. Rifles.

THE SECOND KING OF SIAM TO COL. COLT.

BANGKOK, SIAM, *May* 2, 1859.

. DEAR SIR:—Soon after receiving your letter and beautiful presents, I wrote you expressing my appreciation of
the superior workmanship.

Our country, of course, cannot vie with yours in its
manufactures; but I now send you a sample of the finest
kind of work that is executed in Siam. It is a double
vase of silver, heavily gilt on the inside. The outside is
highly ornamented with gold settings. The gold is the
product of the mines of this kingdom; and, had we
silver-mines, the silver would have been also. The whole
will give you a good idea of the skill and taste of our
people in the manufacture of such articles. I hope it may
be a useful ornament to your table; at all events, please
accept it as a specimen of the product and manufacture
of Siam.

I am only sorry that your already imperishable and

world-wide reputation cannot be augmented by any thing I can say, do, or give.

With best wishes, believe me,

Your friend and well-wisher,

S. P. PAWARENDRRAMESR,

Second King of Siam, &c., &c., &c.

GARIBALDI TO COL. COLT.

GINO, *January* 15, 1860.

NOBLE COLONEL COLT:—As an adopted citizen of the grand Republic, and proud to labor in the all-embracing cause of the peoples, I thankfully, in the name of my country, accept your sympathetic and generous gift.

The arrival of your arms will be hailed among us, not merely as material aid dispatched by a man of heart to a people who fight for their most sacred rights, but as a subsidy of moral potency from the great American nation.

I am, affectionately, your servant,

G. GARIBALDI.

[Translation.]

DUKE OF NEWCASTLE TO COL. COLT.

RICHMOND, VA., *October* 7, 1860.

DEAR COLONEL COLT:—I am sorry to say it will not be possible for the Prince of Wales to accept your kind invitation. After leaving New York we propose to proceed to Albany before going to Boston, so that we shall not be able to visit Hartford.

Had arrangements permitted it, the Prince would have been much pleased to inspect your celebrated establishment.

I am yours very truly,

NEWCASTLE.

February 21, 1861.

My DEAR SIR:—My friend, Col. Helm, Consul-General of the United States, has just transmitted to me, accompanied with a polite letter, two excellent pistols of those recently perfected, that you have the goodness to present to me, and I the pleasure to accept with the highest esteem. Together with that letter, Mr. Helm has sent me one that you addressed to him upon this subject, the highly flattering and benevolent terms of which, in appreciating the services that during my military career I have had the fortune of rendering to my country and queen, place me under obligations, and increase my gratitude.

The arms you present me will be preserved, and placed among mine with the greater pleasure, as coming from a person who, like you, by the force of intelligence and activity, has also rendered a great service to your country, with an invention that has given her a just renown.

Hoping that you will recover your health, and for an opportunity to become personally acquainted with yourself and your worthy family, I am, with consideration and respect, your attentive and obedient servant,

That kisses your hands,

FRANCESCO SERRANO.

[Translation.]

HARTFORD, *March* 15, 1861.

ALL HAIL, COLONEL!—When the Putnam Phalanx visited Mount Vernon, some of its members procured

46

there some portions of wood that was taken, in process of repair, from the mansion of General Washington.

From *this wood* the accompanying cane has been made, by the hand of J. H. Most, Esq. It is surmounted by a head from *wood of the Charter Oak*, and this head is faced with a *shield*, the stripes of which are from wood also of the *Charter Oak*, and the top of which, inlaid with *gold stars*, is from *wood* of the *brave old Ironsides*. The whole forms, as you will at once perceive, a grand historic souvenir, commingling, in a visible and intelligible Trinity, on our own old Tree of Liberty, our Monarch of naval conquests, and, as the shield emblazons with its thirteen stars and stripes, our own immortal Thirteen creative States of the American Republic.

Now, dear Colonel, Mr. Most and myself have thought that, absent as you at present are from your home, and impaired somewhat in the vigor of your limbs, this cane might prove to you an acceptable gift, and furnish assurance that you are kindly remembered.

Accept it, then, with our heartiest compliments, and with our warm hope that it may prove to you, both in a moral and in a physical sense, a stay and a support, warming up your spirits with thoughts of a glorious national Past, and warming up your joints with the healing vigor of renewed and perfected strength.

Dr. Johnson somewhere remarks, that "thousands and millions are of small avail to alleviate the protracted tortures of the gout, to repair the broken organs of sense, or resuscitate the powers of digestion." Now, Colonel, should the accompanying cane, through the associations it inspires, and the support it yields, not only "alleviate," but aid to expel forever from your system every vestige of the dreaded arthritis, it will accomplish its mission to you, there where you are, reposing upon the bosom of the "Queen of the Antilles." That such may be its happy effect, is the sincere wish of Mr. Most, whose cunning skill has fabricated every portion of it, and of one who subscribes himself, as ever,

Very truly yours,

I. W. STUART.

SULTAN ABD EL MEJID TO COL. COLT.

April 10, 1861.

Sultan Abd el Mejid, son of Sultan Mahmoud Khan, may his victories be perpetuated!

The object of this present noble and royal sign, of this illustrious and brilliant world-subduing imperial monogram, is as follows:

The possessor of the present imperial, sublime sign (monogram), Col. Samuel Colt, being an American citizen of talent and great attainment in arts, and moreover entertaining sentiments of a friendly nature for my sublime government, I have conferred on him my imperial decoration of the fifth class, and in testimony of the same I have issued this illustrious *Berat* (diploma) in his favor, in the latter decade of the blessed moon of Ramazan, and the year of the Hejira 1277, in this well-guarded city of Constantine.

EXTRACTS FROM "MINUTES OF EVIDENCE

TAKEN BEFORE THE SELECT COMMITTEE ON SMALL ARMS, MARCH, 1854," ORDERED
BY HOUSE OF COMMONS.

Lieutenant-Colonel Sam. Colt, State of Connecticut,
U. S. A., called in.

1085. "Do you consider that you make your pistols
better by machinery than you could by hand labor?"
—"Most certainly."

1086. "And cheaper, also?"—"Much cheaper."

1087. "Are you familiar with the manufacture of muskets by machinery in America?"—"I am."

1088. "Do you consider that the muskets manufactured by machinery in America are as well fabricated
as the Minié rifle which has been submitted to you?"—
"There is none so badly made at *our* national armories as
the Minié rifle shown to me; that arm would not pass
one of our inspectors."

1089. "Do you consider that the muskets fabricated
by machinery in America are much superior to the Minié
rifle shown to you?"—"Most certainly I do."

1090. "Have you written to the Ordnance to consent,
on certain terms, to supply muskets to the Ordnance?"—
"I have written to the Ordnance my opinion of what the
arms could be made for in England, but not with a view
of taking a contract from that department, but to give
my impression of what the arms could be made for *in
this country*. I have indorsed that opinion by the state-
ment that I would supply the arms at those prices."

1091. "Have you a copy of that letter?"—"Yes; this is the letter which I wrote to them" (*handing in the same*).

1092. "I will read this extract from your letter of the 13th of March. 'So confident am I that this system of manufacturing fire-arms is correct, and the only one by which arms can be made the one like the other, with economy, that I am free to say, what I have before verbally stated, that with one hundred thousand pounds expended in machinery, tools, &c., one million of rifled muskets can be produced at an expense of thirty shillings each; and that, while they will possess the advantage of uniformity of parts, none that are so made will be inferior to the best that can now be found in her Majesty's service for military purposes. I do not want to make a proposition of a contract for the construction of a million of arms *that are not of my own peculiar principle*, while you have men of ability, who, I believe, are fully competent to produce the results I name; yet I am confident that I can produce the one million of arms, if you desire me to do so, at the prices above named, thirty shillings; and I would endeavor to do all the work in this country, unless I should be interrupted by combinations of operatives claiming from me more than the present price of manual labor.'"—"That is all that could be said if I were to talk for a week; that is based on my estimates of what can be done; but if I had to do the work under restrictions that I do not now suffer, it could not be done, or it might not, because people might call upon me for extra service, extra pay, or extra something. If you can insure me the price of labor that I now pay, that result can be produced for that amount of money, with such an amount of outlay."

1102. Lord *Seymour*.] "When did you first commence manufacturing small arms by machinery in America?"— "I commenced many years ago; I commenced in 1836, I think, first to make arms by machinery."

1103. "Since 1836, have you made any great improvements in the machinery?"—"Yes; I have made a great many improvements in the machinery, and every day adds an improvement now."

1104. "When you commenced at first to make arms, was it for the purpose of making arms for the Government of America, or for general sale?"—"For both. At first I intended it for private purposes, but with the hope of supplying Government, as all new mechanics think that Government patronage is valuable to them; it is an advertisement, if nothing else."

1105. "Would the same machinery that answered for the guns made for private purposes, answer also for the musket required for the Government of America?"—"Certainly it would."

1106. "The same machines will make, to a great extent, rifles, or fowling-pieces, or pistols?"—"To a certain extent. Most machinery is competent to many changes, so as to be made applicable to a pistol, to a rifle, to a musket, to a carbine, and to any other arm that you choose to apply it to. You only multiply the simpler engines where you have much to do, otherwise one universal machine would do all that you have to do. If I had one hundred arms to make a year, I should want only three or four machines; but if I had many hundred thousand arms to make, I must have many hundred machines, and then I would make each machine more

stubborn and firm to do this particular work, and in
proportion as it related to the quantity to be made."

1112. "You stated that you could make at the rate of
one hundred thousand muskets a year; but you added, if
no restrictions were put upon you?"—"No more than I
have now."

1113. "Do you mean that the arms when made are
not to be subject to any view?"—"I mean this: give me
ten arms, the best you can pick out, to be models; pick
me ten of the best Minié rifles (do not have them made
on purpose): then pick out the best one, and I will more
nearly resemble the best one with my one hundred thousand a year than the other nine will resemble one
another."

1114. "You have stated that the best is very bad, and
you have offered to make the arm of the Ordnance, but
you say that the arm of the Ordnance is so badly made
that it would not pass view in America?"—"Yes; our
American inspectors are more strict than what I have yet
seen in England; the arms must be more uniform. I do
not say that we can make a single arm so well, or any
better, but I say unqualifiedly this: give one hundred
arms to an American inspector, made at any place you
choose by contract in England, and more would be condemned out of them, even if made expressly, than if you
took the arms made by machinery, which shall be uniform in construction; you give to a man a model, and he
will make them all alike; I can make a bad arm, and
make bad arms all alike."

1115. "Are the Committee to understand that the fault
which you find with the Ordnance is, that one arm is not

like the other?"—"That is the very thing I find fault
with in the arms I have seen here; there is more difference between one and another where they are made by
hand, than there can possibly be when they are made
by machinery. A machine tells better for uniformity
than hand labor does; the eye cannot control the hand
sufficiently to imitate a machine; it is the uniformity of
the work that is wanted. If you give a model, even
though a bad one, and if you instruct your operatives to
attend the machine to make them like the model, they
are all made bad; but if the model is right, the operative
cannot change his machine, or get it out of place, and
that is the effect of machinery on arms, or any other
branch of mechanical industry."

1117. "As you consider the Ordnance arms to be
badly made, do you think the barrel badly made?"—"I
say it is simply bad for uniformity's sake. I say that
you do not make any arms to interchange."

1118. "The fault you find with the arm is, in short,
a want of uniformity of construction?"—"Yes."

1121. "Do you know, practically, whether in the arms
made in America there has been this interchange?"—
"Yes, many thousands; it is the constant habit of Government officers who use my arms in Oregon, or in California, to order one hundred mainsprings, or one thousand
mainsprings, or one thousand hammers, or one thousand
triggers, or one thousand of any thing else applicable to
the most perishable parts of the arm, and I send them;
and one of the commonest operatives or soldiers will take
an ordinary tool, and put them in the place to which
they belong. If the hammer is broken, he will put in a

new one, and so on. Usually it is the custom in our country, where we send out a large number of arms, suppose one thousand, to California or Oregon, for military purposes, to put those parts to which accidents are most likely to occur in the arm-chests. A spring cannot always be relied on so much as a barrel, and we put enough in for immediate repair; and if there is a bad piece of metal thrown in, we supply it. Every thing goes under a particular form of gauge and proof, and an arm is repairable just as well in California as in New York. These remarks are applicable to the national musket, as well as to my own individual arms."

1124. "Their adoption of the manufacture of arms did not arise from the adaptation of machinery to that purpose, but it was before this new machinery was introduced?"—"They began by getting the best tools they could to start with, and continually improving upon them; for instance, improvements in stocking arose from a man who had nothing in the world to do but to attend a common engine-turning lathe; he was turning gun-barrels, and as he came across that part of the barrel projecting for the pan, there was a part that he could not then turn by machinery. Mr. Blanchard said that he could, and, being encouraged, did get an undulating motion in his lathe, and turn that part. That introduced a piece of machinery, and its principle ran through the whole of the Springfield Armory, till they finally produced not only that part better, but they went on till they produced an entire *gun-stock* by machinery. Now it costs less for a stock than it does here for the wood. At that armory this gun-stocking machinery has run on beyond any other branch in every department. That

47

illustrates how far you can go; a man with any quantity
of money can do any thing by machinery."

1137. " Am I right in concluding, that though the
Government of America manufacture themselves a large
amount of muskets, they yet find it good economy also
to purchase of private manufacturers?"—" It may be more
policy than economy."

1138. " Will you explain what you mean by 'more
policy than economy?' is it good policy to maintain the
manufacture in the country?"—" Yes, it is good policy to
maintain a model shop in the country, if nothing more."

1139. " And it is worth while even for the Government of America to make some sacrifice in point of
cheapness, in order to keep up the manufacture in the
country?"—" I think it is good policy of any person who
has to negotiate with another, always to hold within himself the power to produce that which he requires of
another, and it is the same in the case of the Government
of America as in the private transactions of life; they can
produce at the national armories double what they now
do; they do not do so; they wish to encourage the
artisans of the country, and are willing to employ them.
Suppose Government makes half, and contractors the other
half, the result is the same; there is the same inspection;
but for the Government to take the arms harem-scarem,
they would not get the arms they wish for.   Individuals
sometimes, if there is more industry employed, and more
economy employed in labor, may be able to make a little
more profit out of it than the Government does, but that
is the only way; it depends · upon the ability, the
machinery, the capital employed, and the intentions, as to

who wins: Government, who represents the interests of the whole people, or the individual contractors."

1151. "When you first set up your factory here, did you send for machinery from America?"—"No, I first began to buy it here, but it would not do, and I sent to America for it; it was not made for small work. You make a very nice large machine here, but when you come to little details for making fire-arms, I do not know a man competent to do it. Mr. Whitworth comes nearest of those I have met, but he has not made me a good machine yet, nor a perfect machine for my business; he does not make gun machinery; he has not begun to make it. He is, I presume, competent to do it, or any thorough workman, but he has not done it with perfection enough to make guns to interchange."

1152. "Had you much trouble with your men here at first?"—"I had a good deal of trouble to start with. I found great difficulty in this: I brought over first some Americans here, to lead off as my master workmen, but the climate and the habits of the people did not agree with them; they led the jobs. I began here then by employing the highest-priced men that I could find to do different things, but I had to remove the whole of those high-priced men. Then I tried the cheapest I could find, and the more ignorant a man was, the more brains he had for my purpose; and the result was this: I have men now in my employ that I started with at two shillings a day, and in one short year's time I cannot spare them for eight shillings a day; I pay some few of them that, and I would not let them go just now if they required nine shillings. That is the result of employing your men of ignorance. They first come as laborers, at two shillings a day, and, if I find them expert and honest, I employ them as watchmen, or to weigh metal. In a little time, if there be a machine vacant, I put them on it. They would improve from two shillings in the first few months up to four shillings or five shillings, and by and by they become masters. One-half of my masters are Englishmen now, and they fill the vacancies of the Americans who become dissatisfied, from ill-health and from other causes, and go home. The best get eight shillings a day, and at last they become masters too. Do not bring me a man that knows any thing, if you want me to teach him any thing."

1153. Mr. *Walpole.*] "You want good brains, and little knowledge?"—"Yes, I take the raw material."

1154. Mr. *Newdegate.*] "Do you mean that you can get more undivided attention from such men?"—"Yes, the more lessons you can teach them. Take a man here that has been in the habit of rifling a barrel, and he will not do it in any other way than he has been used to do it; but take the raw material, and bring him up to the work (and you have millions of it here), promote the man, and put him in the line of promotion, and he will produce the result you want, and he is elevated."

1170. Mr. *Geach.*] "You have a patent for your pistol; it was taken out in 1836, was it not?"—"Yes; and several for improvements since."

1171. "You were induced to open your establishment for the purpose of carrying the work to perfection, and getting the largest amount of profit that you could out of it?"—"Yes; and I am now proud of the results of my exertions, and can paddle my own canoe."

1172. "It was not, I think, until 1850 that you were enabled to derive the full benefit of the patent?"—"I was not successful here at all; I had to devote all my energies at home."

1173. "I am speaking of America now: until 1849 or 1850, even in America, you were not successful, and you made no profit?"—"It was not profitable; I did not make any money till lately; I made none in America until my arms were employed in the service, by the energy of the people who went first to Florida, next to Texas, in the wars against the Indians, and finally to Mexico, in our war with that country; that completed the reputation of my arm so far as America is concerned. After the Mexican war commenced, its peculiarities being known, the arm, which I was selling to the Government at twenty-five dollars apiece, was sold by individuals and soldiers as high as two hundred dollars to traders; and such was the desire to get those arms, that officers sometimes bought from the traders at those large prices for their own private use, without a knowledge where those arms came from."

1174. "Do you mean your pistol?"—"Yes; my pistols were sold usually for from about seventy-five to one hundred and fifty dollars, and sometimes two hundred dollars each; and the tradesmen resold them to our own officers when the Government paid me only twenty-five dollars."

1175. "In 1847 you had a contract with the Government, had you not?"—"Yes; immediately after the war commenced, then the Government came to me for the arms. The other arms which I made originally got into the service; and when I recommenced my manufacture I advertised in the newspapers for a specimen of my own arm, as I had given my samples and models all away to friends; but I did not find one at the time; and in getting up the new ones I made improvements on the old."

1176. "In 1847 you took a contract for one thousand of those pistols from the Government?"—"Yes."

1177. "For twenty-four thousand dollars?"—"Twenty-eight thousand the first contract, and twenty-five thousand afterwards; the last price was twenty-four dollars each."

1178. "Those one thousand you stated, did you not, cost you twenty-seven thousand dollars to make?"—"I lost by the first thousand; and the reason was, that I had to get machines to make the arms even passable, and I ran my machinery night and day."

1179. "You found that it was some time before you got your machinery to that perfection to which you have raised it now?"—"Yes, it was; I not only improved upon the machinery, but in the model of the arm itself, the same arm for which I contracted, and the result was much better for both the Government and myself."

1189. "Supposing that, instead of private individuals furnishing the money, you had been supplied with the one hundred and fifty thousand dollars that you lost in the first instance by the Government, and they had refused to give any more money, that would have been a failure, would it not?"—"Yes, it was a failure any how; I had to pay thirty thousand dollars out of my own pocket for this failure of private enterprise in making my arms; it is not very creditable to me to have my arms fail under any circumstances, but they did fail; the facts are known to everybody; I have gone to work individually, and I have succeeded by my individual energy and the circumstances of the times."

1214. "What is the price at which you have supplied those arms to the Government?"—"I supply them for any thing I can get."

1215. "What have you supplied them to the Government for?"—"I do not choose to tell you; the Government will answer that question for themselves; I will tell nobody what I supply my arms for. If you want to buy, and say you will buy ten thousand of them, and will give me a fair price, you can have them to-day."

1216. "How are we to estimate the cost of making the machinery by your process, if you will not tell us the price at which you make the article?"—"I will not tell you the price."

1235. "Will you be good enough to state when you commenced the erection of your establishment?"—"The instant I got home, after I had ordered an engine here, which the contractors agreed to furnish in six weeks' time, I made one lot of machinery in America to come here; I put that first lot of machinery in my own American manufactory, and I made a second lot of machinery, and that machinery I also put to work there; this was all done since I left, and since the great World's Fair. After putting the two lots of machinery to work at home, I wrote constantly to know when my engine ordered here would be done, and they said, six weeks after the operatives went to work again. I came here with a third lot of machinery made to start this fabrication, and I waited three months before I got the engine going."

1239. "My object in putting the question is to ascertain how long it is probable the Government here will be in completing a manufactory, by ascertaining how long you had been?"—"That depends on how smart the Government is."

1263. "There is a great difference between your own pattern and the musket on the table, is there not?"—"I should say that I was not competent to make my own, if I could not make such *a thing* as that."

1264. "The first cost of the first fifty thousand would be greater than that of the next fifty thousand, would it not?"—"Yes; I would not undertake to make one hundred and fifty thousand arms without an extra pound, because so much money must be sunk in machines and fixtures to get fairly started."

1265. Colonel *Dunne.*] "In adopting your pistol, have not the Government been satisfied with the shape of your arm, and not bound you to any difficult machinery?" "They have taken it, generally; they admit that my test was pretty good."

1266. "Do not you think they would do the same as to muskets?"—"A musket is an old established thing; it is a thing that has been the rule for ages; but this pistol is newly created; there is nothing that cannot be produced by machinery; if it is necessary to make that hammer (*pointing to the same*) in that particular form, it can be done."

INSTITUTE OF TECHNOLOGY.

FROM the first inception of his great enterprise in the South Meadow, Col. Colt designed to provide for the intellectual improvement and healthy recreation of all in his employ. Soon after the Arms manufacture had become an assured success, he began making provision for a school to popularize science. All through his own lecturing episode, he had observed how readily crowds will gather to attend an experimental lecture. In a similar class-room his own modicum of chemistry had been acquired, and that after his working-hours were over.

He felt that the more mechanics learn of science, the more will they apply of it to their arts, and thus not only render manifold processes cheaper, quicker, simpler, or more perfect, but will strike out other processes as yet unknown. He knew that if working-men were generally educated there would be far more than three learned professions. His own pistol proclaimed what wonderful results may spring from a smattering of science when well applied by a reflecting mind.

In his judgment, then, as a well-wisher to his employés, his best charity was to render science accessible and attractive both to them and to their children.

With this end in view he erected, within a stone's throw of his Arms-factory, CHARTER OAK HALL, an edifice affording apartments for a Reading Room, a Debating Club, and a Library, as well as a Lecture-room, that would accommodate a thousand persons.

But this initiative was no more than a small fraction

48

of his plan. It was an acorn which he meant should
grow to an oak. His aspiration was nothing less than to
rival and surpass the scientific schools which were just
then rising each side of him, in Harvard and Yale.

Accordingly, just one month after the dedication of
Charter Oak Hall, namely, on the sixth of June, 1856,
by his will then made, and by an article here appended,
he bequeathed at least one-fourth of all his wealth for
founding a School of Mechanics and Engineering in
Hartford, and at the same time designated as a site for
its buildings one of the most desirable portions of his
estate.

At that time this endowment was estimated at about
half a million of dollars,—and such were the provisions
of the legacy, that by the present date it would have
amounted to three or four times as much. His desire
was earnest that the institution he intended to found
"might be as beneficent as possible, and might remain
perpetual."

For aid in the organization of this institution he authorized and requested, in 1854, one of the individuals
whom in his will, in 1856, he afterwards named among
the trustees of the institution, to obtain full and reliable accounts of those foreign establishments which had
most features in common with that he had at heart, and
to be prepared to submit a plan in detail.

The munificent bequest stood in the last testament of
Col. Colt nearly three years, and then, owing to a contro-
versy about highways, which had arisen between him
and the city of Hartford, it was revoked and abolished
in 1859. The benevolent purpose that was thus thwarted,
so far as concerned Hartford, Col. Colt still cherished

and designed to fulfil elsewhere. After the breaking out of the Southern rebellion, he intended to superadd military drill to the manual labor feature contemplated in the original scheme, as shadowed forth in his will, and authorized the same individual to include that class of institutions in his inquiries.

No more than two weeks before his decease, he was conversing with the editor of this volume regarding the feasibility of locating the educational institute he had in mind on his estate in East Hartford, and reassured him of his intention to carry out his original purpose. Possibly nothing but the blow which severed him from so many other unfinished plans, prevented his then renewing the princely legacy of other years,—a gift which would have organized the first grand people's college in the world,—established it on a richer fund than any of our scientific schools can as yet boast,—and enshrined its donor's name among that elect few, "who extend the dominion of their bounty beyond the limits of nature, and perpetuate themselves through generations the guardians, the protectors and nourishers of mankind." This might have been—but alas! the truth is—

> "The flighty purpose never is o'ertook
> Unless the deed go with it. What we would do,
> We should do when we would; for this *would* changes,
> And hath abatements and delays as many
> As there are tongues, are hands, are accidents."

EXTRACT FROM THE LAST WILL OF COL. COLT.

"I give, devise, and bequeath to the following persons, viz.: to the Governor of the State of Connecticut for the time being, to the Secretary of the said State for the time being, to the Commissioner of the School Fund

for the time being, the Mayor of the city of Hartford
for the time being, the President of the Connecticut Historical Society for the time being, the Superintendent of
the Colt Patent Fire-arms Manufacturing Company for
the time being, the Engineer of the State for the time
being, the Engineer of the city of Hartford for the time
being, and Isaac W. Stuart, Henry Barnard, Henry C.
Deming, and Richard D. Hubbard, of said city of Hartford, and to their successors as hereinafter provided,
twenty-five hundred shares of the stock of said Colt's
Patent Fire-arms Manufacturing Company; also a certain
square of land in the southeastern corner of the city of
Hartford, within the dykes recently constructed in the
South Meadow, and bounded easterly on Van Dyke
Avenue, southerly on Wawarme Avenue, westerly on
Hendrickson Avenue, and northerly on Wasseek Street;
also a lot of land on the south side of said Wawarme
Avenue and without the dyke, opposite said square of
land above described, and to be of the same size and
shape, and within corresponding lines, together with all
that tract of land which lies between said two squares of
land above described and Connecticut River; the premises
hereby devised are laid down on said map of land here-
unto annexed, to have and to hold to them, the several
persons above described, and their successors, as trustees,
for the following purposes, uses, and trusts, viz.: for the
foundation and establishment of a school or institution for
the instruction and education of young men in practical
mechanics and engineering. Said trustees shall hold the
revenue, income, and profits of the estate above devised
to them for the period of five years from the time of my
decease, as an accumulating fund, and said fund, at the

end of five years from my decease, shall be appropriated
as follows: thirty per cent. for the erection of schools,
workshops, and dwellings on the square above described,
lying within the dykes, for the use of said institution,
said buildings to be substantially erected of stone or
brick; ten per cent. to be set aside as a fund for the support in part of such professors and teachers as may be
employed in said institution; thirty per cent. for the purchase of books and treatises on the subject of mechanics
and engineering, and the necessary tools, implements, and
instruments for giving instruction and employment in practical mechanics and engineering, and the remaining thirty
per cent. for the support of the pupils in the said institution. The estate above devised to said trustees, and the
revenue, income, and profits thereof, shall, after the lapse
of five years from my death, be devoted to the general
purposes of the institution, in such manner as said trustees shall judge to be most promotive of the true intent
and purpose of this devise, which is to afford assistance
to needy and meritorious young men, in the acquirement
of such an education as will enable them to become skilful practical mechanics and engineers.

"The institution hereby founded shall be open to such
young men as said trustees shall see fit to admit; but
my desire is that preference should be given by said
trustees to the sons of the men who are or have been in
my employ, or in the employ of said Colt's Patent Fire-
arms Manufacturing Company, next to inhabitants of the
city or town of Hartford, and next to inhabitants of the
State of Connecticut, and lastly to such other pupils as
said trustees shall see fit to admit, on payment of adequate rates for tuition and instruction, and so forth.

All pupils belonging to this State, who shall be admitted
by said trustees, shall, if unable to pay the expenses of
their education in said institution, give such obligation,
bond, or security as said trustees shall see fit to require,
for the repayment to the said trustees of the expenses of
said education, whenever, after leaving said institution,
they shall be able so to do. All pupils from this State
who are of sufficient ability, and all pupils from other
States, shall pay such rates for tuition as said trustees
shall prescribe. The pupils in said institution shall be
dressed in appropriate uniform, and their time, with due
and reasonable allowance for healthful sports and recreation, shall be equally divided between theoretical studies
in mechanics and engineering, and practical employment
in the workshop.

"Said institution shall be under the care and management of said trustees; five of whom shall be a quorum
for the transaction of business; and in case death, resig-
nation, or inability or disability of either or any of the
four trustees last named, shall occasion a vacancy in said
board, the Mayor of the city of Hartford for the time
being shall appoint a trustee to supply said vacancy temporarily, and the electors of the city of Hartford are em-
powered and requested, at their annual or other legal
meeting, to elect by vote a permanent trustee to fill such
vacancy; but every such person, so appointed or elected,
shall be an elector of the town of Hartford. My oldest
male relative, who for the time being is next of kin and
bears my family name, and is twenty-one years of age at
least, shall be entitled to be President of said Board of
Trustees.

"It is my desire that said trustees shall be incorporated

as a body corporate and politic, for the purposes, and under the organization, and with the order of succession above prescribed, to the end that the institution hereby intended to be founded shall be rendered as beneficent as possible, and remain perpetual." * * * (*June* 9, 1856.)

* * * "and in lieu thereof, I give and bequeath said five hundred shares of stock to the trustees named in said will for founding a school for practical mechanics and engineers, subject to the uses and trusts created in said will for that purpose."—(*Codicil, January* 12, 1858.)

"I hereby cancel and revoke the devise and legacy heretofore made, in and by said original will and codicil, to the Governor of the State of Connecticut and others as trustees, for the purpose of founding a school for the education of practical mechanics and engineers. My design is to abolish said devise and bequest."—(*Codicil of February* 2, 1859.)

The following resolutions were passed by the Putnam Phalanx, at a meeting held January 14th, 1862:—

WHEREAS, It has pleased Almighty God, in His wise providence, to remove from our ranks, and to take from the world, our late esteemed and highly honored fellow-member, Col. SAMUEL COLT—therefore,

Resolved, That in the death of Col. Colt, in the prime of his manhood and in the midst of his usefulness, our Battalion has sustained an irreparable loss, which we sincerely deplore.

Resolved, That Hartford may well weep over the early fall of her honored son, to whom she owes much of her prosperity and glory, and the fruits of whose genius and industry will remain to bless our city through unborn generations.

Resolved, That the inventions of Col. Colt, combined with his wonderful executive power, as exhibited in his improved fire-arms, submarine batteries, and the like, gave immense importance to his operations in business, and render his early death a calamity to our country.

Resolved, That though young in years, Col. Colt was old in deeds, and that "we live in deeds, not years; in thoughts, not breath; in feelings, not in figures on a dial. We should count life by heart-throb—he most lives who thinks most, feels the noblest, acts the best."

Resolved, That while we deeply sympathize with the bereaved family of our departed fellow-member, a copy of this preamble and these resolutions be presented to them, as a faint expression of the deep sorrow that fills all our hearts.

Resolved, That in token of respect for the memory of Col. Colt, we will wear the usual badge of mourning for one year.

Resolved, That these, our expressions, in view of the sad event we mourn, be published in the city papers.

The following resolutions were passed at the meeting of the workmen on Saturday:—

WHEREAS, It has pleased the All-Wise Disposer of Events to remove from among us one upon whose energy, enterprise, and public spirit, we have been accustomed to depend for the support of ourselves and families, and one who has ever stood by the interests of working-men through all the great financial struggles which have from time to time agitated the country in years past; therefore,

Resolved, That in the death of Col. SAMUEL COLT, the mechanics and laboring men have lost a friend who could sympathize with them in their efforts, their toils, and their trials, and one who could appreciate genius and talent, however humble the garb under which it was found.

Resolved, That by this sad event we, his employés, are called upon to mourn the loss of one, the conceptions of whose powerful mind, and the execution of whose gigantic plans, have contributed largely to the happiness of ourselves and families, by furnishing us the means of an honest livelihood; and it is our fervent prayer that his mantle may fall upon those who will carry out the liberal policy which he has ever pursued with such marked success.

Resolved, That our sympathies are hereby respectfully tendered to his family and friends, whom we recognize as having lost by this sad bereavement a devoted husband, a fond and loving father, and a warm-hearted and genial friend.

Resolved, That we will attend the funeral in a body, and take such part in the service of the occasion as may be deemed proper; thus showing, as far as in us lies, that respect for him in death which in life he so justly merited.

Resolved, That the Secretary be instructed to present a copy of these resolutions to the family of the deceased, and also to procure their publication in the daily papers.

49

CONCLUSION.

CONCLUSION.

Few inventors realize the ideals they originate, or derive pecuniary advantage from the devices of their brains. Blanchard, who, aside from his gun-stock miracle, has obtained three and twenty other patents, has from few of them accumulated any considerable profits. While enriching the world they impoverish themselves. Their fame is proverbially "an estate which they inherit after death." Moses, who discovers the land of promise, dies in the wilderness. Joshua and the next generation pass over Jordan. But in reviewing the career of Col. Colt, we know not which most to admire,—the progressive development in his own models of his boyish idea, or his success in flashing it abroad through the world like the lightning which shines across the firmament from east to west. Bacon could not hope that his speculations would be appreciated till "after some ages." When past threescore he wrote, *Mihi satis fuerit sevisse posteris.* On the other hand, within a dozen years after Colt commenced making pistols in Hartford, he had reared the colossal Arms Factory, turned out a million arms, gained a clear profit of three millions of dollars, and added five thousand to the population of that city of his birth, and

where, notwithstanding tempting proposals elsewhere, he
preferred to build his palace of industry. Since his patent
expired, his ideas furnish occupation to more out of Hartford than in it. No matter, then, though his manufactory be burned to ashes, and not one stone be left on
another. His invention would rise as from the pyre of
the Phœnix, and clothe itself in a new, perhaps a more
noble embodiment. It is a true cosmopolite.

Though he was cut down in the midst of his days,
his epitaph may well be—

> " He in our wonder and astonishment
> Has built himself a live-long monument."

Nor should it be forgotten—though it often is—that he
who gives men employment, and hence feeds the labor-
loving, is a more genuine benefactor than he who enables
them to eat the bread of idleness. While the one
bestows " the glorious privilege of being independent," the
other teaches men to crook the pregnant hinges of the
knee when thrift will follow fawning. Hence the stern
Spartan's answer to a beggar was: He who first gave
you alms robbed you of manliness, and so took more
than he gave.

By obtaining an act of incorporation, Col. Colt had
made it as sure as he could that his inventions would
have free course after his death. Nor is it easy to find
any thing to cavil at in the daily working of the Arms
Factory.

After all, his friends cannot but imagine to themselves
what marvels he would have wrought, had he fulfilled
the three-score and ten years which make up the full
age of man. If he brought so much to pass in ten years,

starting with nothing, what would he not have done in
twenty years more, starting with a capital of millions?
When men achieve great and sudden successes, their
heads are apt to be turned. Howitt says of the man
who found the great nugget which weighed twenty-eight
pounds at the Bendigo diggings in Australia:—"He soon
began to drink; got a horse and rode all about, generally at full gallop, and when·he met people, called out
to inquire if they knew who he was, and then kindly
informed them that 'he was the bloody wretch that had
found the nugget.' At last he rode full speed against
a tree, and nearly knocked his brains out." And he
was a type of his class. Yet, dazzling as was the prize
of Col. Colt, his head was not turned. On the contrary,
all successes were to him but stepping-stones on which
he could mount to a loftier plane of enterprise and
accomplishment. His heart was fixed, not on what he
had done, but on what remained still to *be* done.

His willow-workers are assured that he would have
opened so many a market for their furniture, so neat-
fingered and fairy-like, that their numbers would be
reckoned by hundreds rather than by scores. The Artesian well which he left half bored he would have sunk
so deep that its stream—rushing up from the recesses of
the earth, in the miracle of water running up hill, and
soaring in a sheaf of foam high as an Icelandic Geyser
as it fell—would fill his fish-ponds, as well as diffuse comfort, cleanliness, and safety from fire, through a thousand
dwellings.

He had extended his products from pistols to rifles,
and from rifles to muskets. So he would have added
the fabrication of great guns, and, rivalling South Boston

and Pittsburg, would have evinced that, in constructing the smallest of arms, he had learned lessons not without use in forming those most gigantic.

His wealth had already overflowed in house and grounds,—pictures and statuary,—garden with deer, pea-cocks, and swans,—a green-house two thousand one hun-dred and thirty-four feet in length, and the crystal palace of a conservatory, marble-floored, with fountains and chandeliers, "a stately pleasure dome,"—a group of luxu-ries which has seldom been surpassed.   In this direction he would have gone on as he had begun.

He had at heart the mental culture of working-men. Accordingly, he had at one time made provision in his will for establishing in Hartford, on a princely foun-dation, an institution which would embrace courses of lec-tures, scientific and literary, open to all; an ample libra-ry, and a polytechnic school, where whoso would might learn the history of every art, together with its relations to science, its desiderata, and the directions in which improvements are to be sought.   Though the bequest, larger than that of Smithson, which he had made for this end,—owing to an unfortunate misunderstanding,— was expunged from his last testament, yet, had his life been prolonged, it would have been inserted again, not only for the benefit of his native city and State, but of the whole country, inasmuch as he was clearly one of those noblemen of nature, described by Burke as "of an insa-tiable benevolence, who, not contented with reigning in the dispensation of happiness during the contracted term of human life, strain, with all the reachings and graspings of a vivacious mind, to extend the dominion of their bounty beyond the limits of nature, and to perpetuate themselves through generations of generations, the guardians, the protectors, and the nourishers of mankind." Of this sort was the stream of his character, however counter to it some of his impulsive eddies may have dashed.

It is because such as these were the anticipations of men regarding him, that they said at his fall,—"The mainspring is broken;" and the eyes of all who knew him will always miss something whenever they look toward the scene of his triumphs. Worthy men—of rare inventive and administrative talent—succeed him; but they cannot replace him.

> "As in a theatre the eyes of men,
> After a well-graced actor leaves the stage,
> Are idly bent on him who enters next,
> Thinking his prattle to be tedious."

Yet great as were Col. Colt's achievements, and however much greater they might have been, had his been length of days, they were, and they would have been, as nothing to his ideal. Where are the forms the sculptor's soul has seized? In *him* alone. No truth is plainer, in the biographies of those whose genius has changed the face of the world, than that they have surpassed others not so much in what they have done, as in what they have aspired to do. One of the most perfect of historians, Arnold, says: "There floats before me an image of power and beauty in history which I cannot in any way realize, and which often tempts me to throw all I have written into the fire." In like manner, we may figure to ourselves young Colt, on the day when the first ideal of a matchless fire-arm flitted before him, on his voyage to Calcutta, as doubling the Cape of *Good Hope* in a higher sense than the rest of his ship's company. He

50

then rounded that Cape of Hope which to him was prophetic of the better Indies. Even then, it may be, he
said to himself: "This ribbon-block, in which I have pierced
six holes,—and which to the sailors seems the play-
thing of an idler,—shall work as great wonders as I have
read of in the Arabian Nights. It is my lamp of Alad-
din. It shall play no mean part in that art of war,
which, wherever men dare assert freedom, or seek to
establish despotism, is the art of arts. My handiwork
shall be counted the one thing needful wherever the hand
of man is against his fellow, or against wild beasts.
What then is my secret less than the potentiality of
growing rich beyond the dreams of avarice? In my
native town, which I am now homesick to behold, I will
rear me a palace,—beautiful for situation, and within
which the great masters of every fine art shall clothe the
palpable and the familiar with golden exhalations.   I
will surround it with a paradise of flowers, fruits, fountains, and animals. There will I enthrone the idol of
my heart, and subject all the shows of things to the
desires of her mind. Thence will I overlook a new
quarter of the city rescued from flood by my Dutch
mound, and filled with my retainers. Some millions will
I dedicate to a school, where art and science shall unfold
their mutual relations more harmoniously than ever before.   There theory and practice, each supplying what
the other lacks,—shall embrace each other. There, whatever seed of invention lies latent in me, or in those of
my friends whose minds are capacious of such things,
shall take root downward and bear fruit upward. The
submarine explosives I used to experiment with on Ware
Pond, I will so elaborate as to afford to all my coun-

try's harbors a defence, cheap yet impregnable against all the navies in the world. Through my patronage, expedients to render labor of all sorts of more power to secure sublimest results by simplest means shall multiply. The useful arts—one and all—shall become so interpenetrated with science, that they shall slough off their meanness, and be recognized as equal to their sisters who arrogate the titles of 'fine' and 'liberal' as all their own. Labor, through becoming intellectual, shall assume a new dignity."

Far higher than this he may have climbed in the air-castles of childhood dreams, which sheening far celestial seem to be,—*aliquid immensum infinitumque.*

Man proposes, but God disposes,—and God only is great.

> " With noiseless step death steals on man ;
> No plea, no prayer delivers him ;
> From midst of life's unfinished plan,
> With sudden hand it severs him."

> "No longer seek his merits to disclose,
> Or draw his frailties from their dread abode ;
> There they alike in trembling hope repose,
> The bosom of his Father and his God."

Col. Colt died on the morning of Friday, January 10th, 1862, and the funeral services were performed at his late residence on the afternoon of Tuesday, the 14th, with such attendance of friends and demonstrations of public sorrow and sympathy as are rarely witnessed on the death of a private citizen.

His body was borne out to return no more by a delegation from his warm personal friends and neighbors,—through crowded lines composed of the Putnam Phalanx, and of workmen in the Armory to the number of fifteen hundred,—across the lawn, to the grove which he had planted, and where two of his children, who had died before him, were already sleeping, and to the same resting-place was he consigned.

His grave is covered by a plain slab of white marble, for which the bereaved wife and mother dictated this simple but expressive epitaph:—

KINDEST HUSBAND, FATHER, FRIEND, ADIEU.

Any other monument, on these grounds, within his Dyke, or within sight of the Armory, and any other inscription, to be read so near the home where he had garnered up his heart's best affections and aspirations, and so full of the tokens of considerate and lavish kindness, would be not only unnecessary, but almost obtrusive. But in the new cemetery recently laid out on Cedar Hill, on an eminence which commands a view of the city and the valley which he loved so well, Mrs. Colt has selected a family burial-place, and there, among other

BORN
JULY 19, 1814.
DIED
JAN. 10, 1862.

COLT

memorials of his towns-people and friends, will rise a monument of enduring material and massive proportions.

On the northern front of the base is engraved the name of

SAMUEL COLT:

and above, on the face of the die,

BORN
July 19th, 1814,
DIED
January 10th, 1862:

and over these inscriptions, on the column base, are set the Colt family arms.

On the eastern side of the base is the following inscription :—

IN MEMORY OF A BELOVED HUSBAND AND OF OUR DEAR
CHILDREN THIS MONUMENT IS ERECTED.

The monolithic sub-base, of gray granite, was quarried at Millstone Point, near New London, Connecticut; is twelve feet square by two feet thick, weighing twenty-five tons. The superstructure entire is of highly wrought rose-colored sienite, from the celebrated quarries at Peterhead, in the north of Scotland, and consists of the following parts :—

The BASE, nine feet six inches square by fourteen inches thick.

The DOUBLE PLINTH, rich moulded, seven feet seven inches in diameter by two feet five inches in thickness.

The DIE, or cubical part of the pedestal, three feet eight inches square, the faces being sunk sufficiently to work the Egyptain torus, or beard, upon the arrises or edges.

The Cornice, six feet four inches square by thirteen inches in thickness, in one piece.

The Column Base, four feet five inches square by twenty-two inches thick.

The Main Shaft, thirteen feet one inch in rise, circular, having a diameter at foot of thirty-six inches.

The Capital, four feet five inches diameter by two feet eight in rise, richly carved and polished.

The Statue surmounting the capital is bronze, representing the Angel of the Last Judgment, modelled by the sculptor Randolph Rogers, at Rome, expressly for this work, and was put in bronze at the Royal Foundry in Munich.

The style of architecture is Egyptian, the main shaft and capital having been modelled from the examples and suggestions found among the ruins of the grand temple at Karnac; and the four smaller columns, standing at the angles of the die, and supporting the cornice, from an ancient temple at Luxor.

The columns are ornamented with the pyramidal reliefs and horizontal sinkings peculiar to Egyptian architecture.

The capital bears the lotus flower, from which it was originally designed, even as the Corinthian in later days found its origin in the acanthus.

The Winged Globe variously composed, but generally of the disk of the sun, emblematic of celestial light, flanked by the inner wings of the sacred beetle, symbolic of the creating spirit; or with two asps depending, symbolic of sovereignty and eternity. It was a favorite symbol among the ancient Egyptians, and not only appears upon the gateways of their cities, but upon temples and tombs alike, as the heraldic arms of their country. Its general meaning may be characterized as a symbolic figure of the Providence of God overshadowing the land of the Nile; and, in its larger application, of the protecting genius of the world.

These symbols are the more interesting when taken in connection with the ideal figure which crowns the monument, showing the progress of religious faith, and the triumph of Christianity over error and superstition, "in abolishing death, and bringing life and immortality to light." In this understanding of the figure, towering from its position and its own proportions above the other monuments of the cemetery, it becomes a most significant symbol of the higher meaning of this consecrated ground.

The whole height of the monument is forty feet, and the general design, composition, and construction are all in complete harmony, and consistent with the materials used; while its massive proportions correspond with the power and energy of the man whose name it will bear for all time to come.

The monument, as a whole, was designed and constructed by James G. Batterson, Architect and Sculptor, of Hartford.

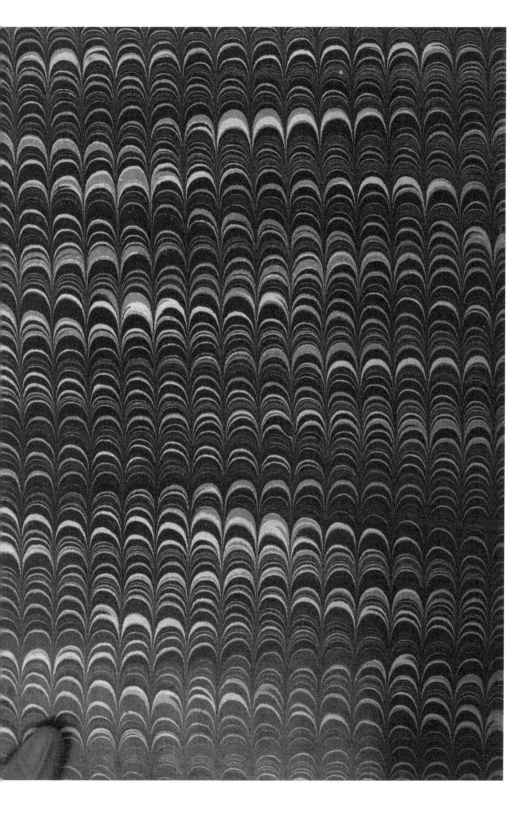